MORSE THEORY

Smooth and Discrete

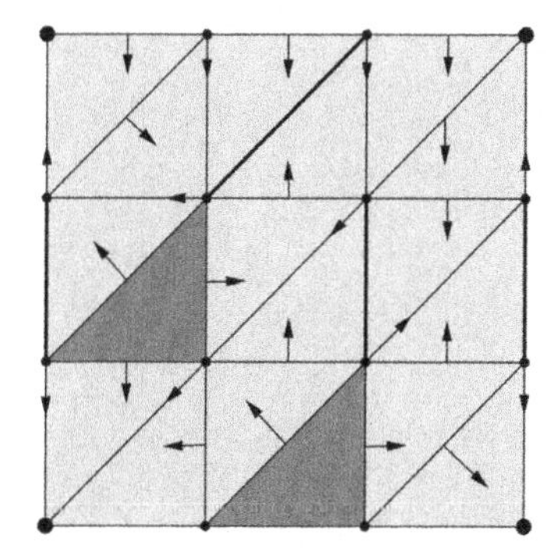

MORSE THEORY

Smooth and Discrete

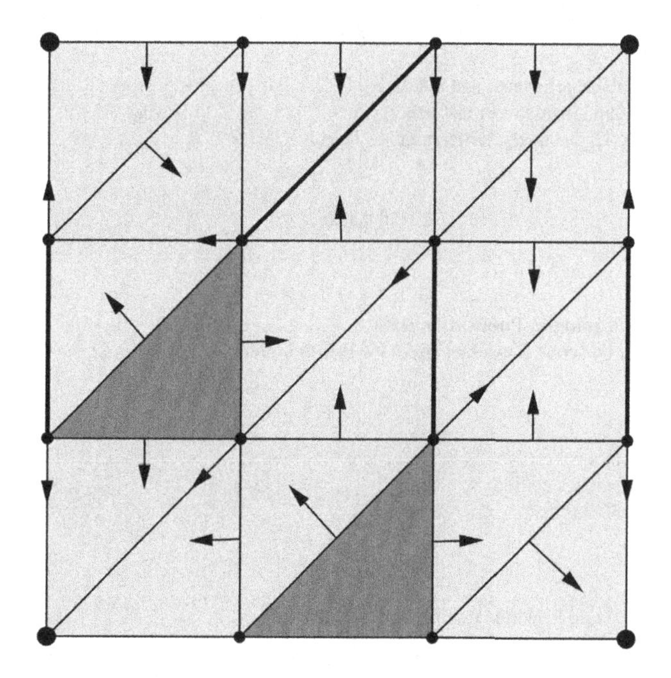

Kevin P. Knudson
University of Florida, USA

NEW JERSEY • LONDON • SINGAPORE • BEIJING • SHANGHAI • HONG KONG • TAIPEI • CHENNAI

Published by

World Scientific Publishing Co. Pte. Ltd.
5 Toh Tuck Link, Singapore 596224
USA office: 27 Warren Street, Suite 401-402, Hackensack, NJ 07601
UK office: 57 Shelton Street, Covent Garden, London WC2H 9HE

Library of Congress Cataloging-in-Publication Data
Knudson, Kevin P. (Kevin Patrick), 1969–
Morse theory : smooth and discrete / by Kevin P. Knudson (University of Florida, USA).
pages cm
Includes bibliographical references and index.
ISBN 978-9814630962 (hardcover : alk. paper)
1. Homotopy theory. 2. Geometry, Differential. I. Title.
QA611.K575 2015
514'.24--dc23
2015008832

British Library Cataloguing-in-Publication Data
A catalogue record for this book is available from the British Library.

First publishd 2015 (Hard cover)
Reprinted 2015 (in paperback edition)
ISBN 978-981-4740-56-2 (pbk)

Printed in Singapore

To Ellen and Gus, for making three a magic number

Preface

Like most mathematicians of my generation (and those before us), I first learned Morse theory via the "bible," John Milnor's exquisite text *Morse theory.* To this day it remains, like many of Milnor's books, a masterwork of clarity and exposition. The only drawback, if one wants to call it that, is that the book contains only about 40 pages of actual Morse theory; from there Milnor goes on to prove what he calls the Fundamental Theorem of Morse Theory, which computes the homotopy type of the space $\Omega(M; p, q)$ of paths from p to q in the complete Riemannian manifold M. This is a beautiful result, to be sure, and allows one to deduce the Bott Periodicity Theorem, for example, but it may be a bit much for a first introduction to the subject, especially for an advanced undergraduate or beginning graduate student.

A more gentle, yet no less elegant, exposition may be found in Yukio Matsumoto's *An Introduction to Morse Theory.* Matsumoto's text contains a wealth of detailed examples and beautifully illustrates the fundamental ideas of the subject in a manner that is accessible to students equipped with a solid understanding of basic topology and advanced calculus. He concludes with some applications to the Kirby calculus of handlebodies and links.

In light of this, I have not tried to reinvent the wheel. The first part of this book contains an introduction to Morse theory via a presentation following the texts of Milnor and Matsumoto (with occasional input from Milnor's book *Lectures on the h-cobordism Theorem* and a few other sources). Many of the proofs are standard parts of the literature by now; if they are not in Erdös's *Book* they are as close as we can get to those ideals. The topics were chosen with an eye toward the second part of the book, where we meet their modern counterparts in discrete Morse theory. Moreover, it should be mostly accessible to those with a knowledge of topology and multivariable calculus.

In the mid-1990's, Robin Forman developed a version of Morse theory on arbitrary CW-complexes. Dubbed *discrete Morse theory,* this combinatorial definition of a Morse function allows one to deduce for general complexes much of what smooth Morse functions say about the topology of a smooth manifold. The definitions are quite simple, and one quickly reduces them to a study of the associated gradient vector field. This latter object has a well-studied combinatorial interpretation, that of an *acyclic matching* on a certain directed graph. Because of this correspondence, one is able to use graph theory to construct these vector fields and to develop efficient algorithms for doing so. This leads to many interesting applications.

The book is organized as follows. Chapters 1 through 4 cover the basics of Morse theory, from the first definitions (Morse function, nondegenerate critical point, etc.), to the Morse Lemma and associated gradients, to handle decompositions of manifolds, and finally to the Morse complex for computing homology. Along the way, I have tried to include many examples to clarify and illustrate; the reader is invited to supply others. For some of the more technical results, such as handle canceling or the construction of the differential in the Morse complex, I have chosen not to give complete proofs. Rather, sketches are given which should be sufficiently convincing, and the corresponding results in the discrete theory are then proved in full detail.

Chapter 5 is a hybrid, properly belonging to neither Part I nor Part II (but landing in the former–a choice had to be made). It discusses various piecewise linear approaches to Morse theory, which may be viewed as first attempts at a Morse theory on simplicial complexes. The first section presents the critical point theory on polyhedra embedded in $\mathbb{R}^3$ as developed by Banchoff. In some sense, this idea is the foundation for much of the discrete Morse theory in subsequent chapters.

Part II is devoted to Forman's discrete Morse theory. Beginning with the basic definitions, the analogues of the theorems of Part I are proved in this more general setting. One notices that many of the theorems whose proofs are highly technical in the smooth case become much simpler in the discrete setting. With these notions in hand, the final two chapters of the book present algorithms for constructing discrete Morse functions and applications of these ideas to various fields in mathematics and data analysis.

I have included some exercises to aid an instructor in using this book as a text for a course on this material. Some of them are routine; others are more challenging. One problem is that it is somewhat difficult to generate exercises for the last two chapters. Indeed, many of the most interesting applications of discrete Morse theory involve the implementation of algorithms for the construction of the associated vector fields. The reader is invited to (a) write implementations or find them online, and (b) try them out on interesting data sets. One such implementation, of Algorithm 8.22, may be found online at the web address indicated in the text.

Finally, there are two appendices, one on smooth manifolds and one on cell complexes, for quick reference. They are no substitute for texts in manifold theory and algebraic topology, but they do highlight the concepts used in the body of the book. I hope the reader finds them useful.

K. P. Knudson
Gainesville, FL
February 2015

Acknowledgments

As anyone who has written a book knows, it is an exercise in frustration, perseverence, despair, and joy. This is my second book, though, so it must be worth it somehow, and I would like to thank a few individuals and institutions who have made this possible.

I have had the privilege of learning from many excellent topologists over the years. In particular, I am indebted to Professors Peter Fletcher, Richard Hain, and John Harer for leading me down this path. My interest in discrete Morse theory was sparked in 2004 during a workshop on Computational Topology organized by John Harer and Herbert Edelsbrunner at the Institute for Mathematics and its Applications at the University of Minnesota. It was at that same workshop where I first met my collaborators, Henry King and Neža Mramor–Kosta; I am grateful for their good humor and even better ideas. I would also like to thank Uli Bauer and Dmitry Kozlov for many enlightening conversations about discrete Morse theory over the years and I hope we have many more in the years to come.

I thank Rochelle Kronzek at World Scientific for encouraging me to bring this project to fruition. I had been kicking this idea around for some time, but I needed a final nudge. Thanks also to Ms. E.H. Chionh for her editorial assistance.

This book was written while I was on research leave from my faculty position at the University of Florida. I thank our provost, Dr. Joe Glover, for allowing me this "time off for good behavior" as he put it. I spent a good deal of time writing at Volta Coffee, Tea & Chocolate in downtown Gainesville; their beautifully crafted lattes sustained me.

Finally, I am supremely fortunate to have a wonderful family. My wife, Ellen, and my son, Gus, are wildly creative people who fill me with pride every day. Ellen makes beautiful artist's books; Gus writes beautiful music. I only hope I can make mathematics seem half as lovely.

Contents

Preface vii

Acknowledgments ix

Smooth Morse Theory 1

1. First Steps 3
   - 1.1 Introduction . . . 3
   - 1.2 Critical points . . . 5
   - 1.3 Morse functions on surfaces . . . 8
   - 1.4 Exercises . . . 11

2. Fundamental Results in Morse Theory 13
   - 2.1 The Morse Lemma . . . 13
   - 2.2 Existence of Morse functions . . . 17
   - 2.3 Gradient-like vector fields . . . 21
   - 2.4 Integral curves . . . 23
   - 2.5 Exercises . . . 25

3. Topological Consequences 27
   - 3.1 Handle decompositions of manifolds . . . 27
   - 3.2 Examples of handlebody decompositions . . . 34
   - 3.3 Sliding and canceling handles . . . 39
   - 3.4 Exercises . . . 49

4. Homology 51
   - 4.1 Cellular homology . . . 51
   - 4.2 The Morse complex . . . 54
   - 4.3 The Morse inequalities . . . 59

4.4 Poincaré duality 60
4.5 Exercises 64

5. Piecewise Linear Morse Theory 67

5.1 Critical points on embedded polyhedral surfaces 67
5.2 Computational Morse theory 70
5.3 Another version of PL-Morse theory 79
5.4 Exercises 82

Discrete Morse Theory **83**

6. First Steps 85

6.1 Basic definitions 85
6.2 Simplicial collapses 90
6.3 Discrete vector fields 92
6.4 Dynamics of discrete vector fields 100
6.5 A graph-theoretic point of view 101
6.6 Exercises 103

7. Topological Consequences 105

7.1 Homotopy type 105
7.2 Sphere theorems 107
7.3 Canceling critical cells 109
7.4 Homology 111
7.5 Comparison with smooth Morse theory 121
7.6 Exercises 122

8. Algorithms 125

8.1 First case: 2-dimensional complexes 125
8.2 General n-complexes 127
8.3 From point data to discrete Morse functions 131
8.4 Optimality 135
8.5 Exercises 139

9. Applications 141

9.1 Combinatorial applications 141
9.2 Geographic data 146
9.3 Evasiveness 148
9.4 Algebraic discrete Morse theory 152
9.5 Homology reduction algorithm 156
9.6 Exercises 161

Appendix A Smooth Manifolds 163

Appendix B Cell Complexes 169

Bibliography 175

Index 179

PART 1
Smooth Morse Theory

Chapter 1

First Steps

1.1 Introduction

If you have ever taken a hike over hilly terrain, then you already understand the basic notions of Morse theory. For example, you know that the fastest way up a hill is to follow the gradient, which gives the direction of fastest increase in elevation. When walking you encounter three basic kinds of critical points:

(1) the bottom of a hollow, corresponding to a local minimum of the elevation function;
(2) the top of a hill, corresponding to a local maximum of the elevation function;
(3) a low point on a ridge, corresponding to a saddle point at which moving in one direction leads to an elevation increase while moving in the orthogonal direction leads to an elevation decrease.

As a specific example, consider the image in Figure 1.1, which shows the profile of Pilot Mountain in North Carolina. We immediately see two maxima and one saddle point lying between them. The summit of the knob on the right is the absolute maximum, while the lower peak on the left is merely a local maximum.

The topographical map of the region around the summit is shown in Figure 1.2. The curves are the level curves of the elevation function. If we think of this function as a map $f : D \to \mathbb{R}$, where D is the square shown, then ∇f is orthogonal to the level curves at each point. Moreover, we can easily find the critical points of f corresponding to the two maxima and the one saddle point.

Now imagine you are hiking on the mountain and you are at one of the maxima (actually, you couldn't be on the summit of the knob as it is closed to hiking and climbing). Call this point p. From your point of view, the mountain looks much like an inverted bowl. In mathematical terms this means that we can choose local coordinates (x, y) around the maximum so that the function f takes the form

$$f(x, y) = f(p) - x^2 - y^2,$$

which is simply an inverted paraboloid. Similarly, if q is the saddle point, your view from q would look much like a saddle with two choices for moving down the

Fig. 1.1 Pilot Mountain, NC

mountain, opposite from each other, and two choices for moving up the mountain, orthogonal to the down direction. In functional notation, we may choose local coordinates (x, y) so that near q, we have

$$f(x, y) = f(q) - x^2 + y^2.$$

Finally, if you are at point s at the bottom of a hollow (and there are no obvious hollows in this map), it looks as if you are at the bottom of a bowl, and locally we can write

$$f(x, y) = f(s) + x^2 + y^2.$$

Observe that in each case, the local representation is a quadratic function in the coordinates with 0, 1, and 2 negative signs in the cases local minimum, saddle, and local maximum, respectively. As we shall see, these numbers are independent of the local representation and therefore give a well-defined quantity called the *index* of the critical point. Moreover, these critical points are *nondegenerate* in a very specific sense to be defined below. Given a smooth function with only nondegenerate critical points, we shall show that it is always possible to represent the function f locally by such a quadratic function.

Smooth Morse theory tells us that if one has such a function f on a smooth manifold M, then the global topology of M is determined by the critical points of f in a very precise way. Indeed, we will be able to use information about the various critical points to construct a cell complex having the homotopy type of M and also to construct a chain complex computing its homology. This is a remarkable result as there is no *a priori* reason to believe that isolated local information can be assembled to tell us so much about M.

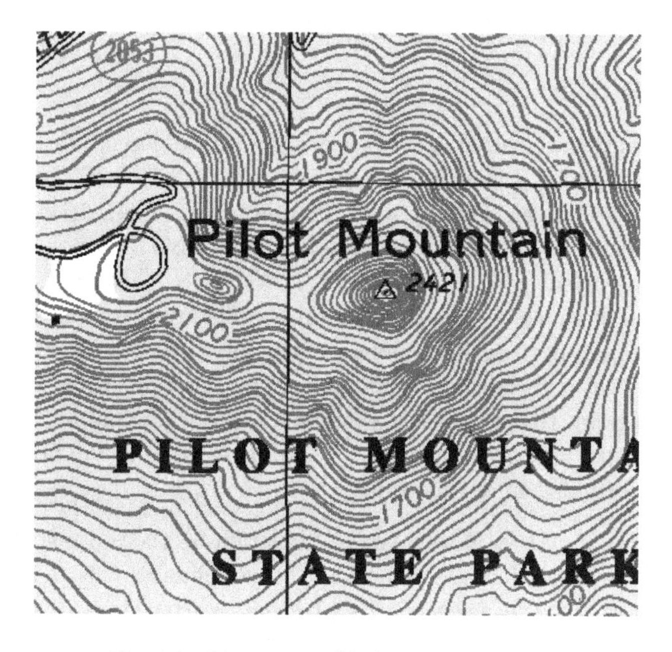

Fig. 1.2 The topographical map of Pilot Mountain

1.2 Critical points

Let M be a smooth n-dimensional manifold and suppose that $f : M \to \mathbb{R}$ is a smooth function (see Appendix A for the relevant definitions).

Definition 1.1. Let p be a point on M and choose local coordinates $(x_1, \ldots, x_n)$ of a neighborhood of p such that p is the origin. Then p is a *critical point* of f if

$$\frac{\partial f}{\partial x_i}(p) = 0, i = 1, \ldots, n.$$

For surfaces embedded in $\mathbb{R}^3$ this simply means that the tangent plane to M at p is horizontal.

Example 1.2. Consider $f(x, y) = x^2 + \cos y$ on $M = \mathbb{R}^2$. The origin is a critical point of f:

$$\frac{\partial f}{\partial x} = 2x \quad \text{and} \quad \frac{\partial f}{\partial y} = -\sin y$$

and both of these vanish at $(0, 0)$.

Example 1.3. Consider $g(x, y) = x^3 - 3xy^2$ on $M = \mathbb{R}^2$. The origin is also a critical point of g:

$$\frac{\partial g}{\partial x} = 3x^2 - 3y^2 \quad \text{and} \quad \frac{\partial g}{\partial y} = -6xy$$

and these vanish at $(0, 0)$.

The critical points in the above examples are of a very different nature, however. Since $\cos y \approx 1 - y^2/2$ for y near 0, we see that $f(x, y) \approx 1 + x^2 - y^2/2$ near the origin and therefore that the origin is a saddle point of f. The critical point of g, however, is degenerate in the sense that there are multiple directions of increase and decrease at the origin.

An even simpler example of this phenomenon is familiar to first-semester calculus students when comparing the critical points of the functions $a(x) = x^2$ and $b(x) = x^3$. In the former case, $x = 0$ is a minimum of the function, while in the latter $x = 0$ is a critical point that is neither a maximum nor a minimum. The primary difference is that, at $x = 0$, we have $a''(0) = 2$, while $b''(0) = 0$. We now introduce the more general formulation of this concept.

Definition 1.4. Let $f : M \to \mathbb{R}$ be a smooth function on the n-dimensional manifold M and let p be a point on M. Choose local coordinates $(x_1, \ldots, x_n)$ near p. The *Hessian* of f at p is the $n \times n$ matrix of second partial derivatives of f. We denote by $H_f(p)$ its determinant

$$H_f(p) = \det\left(\frac{\partial^2 f}{\partial x_i \partial x_j}(p)\right).$$

Remark 1.5. Suppose that $(z_1, \ldots, z_n)$ is another set of coordinates near p. Then we have

$$\left(\frac{\partial^2 f}{\partial x_i \partial x_j}\right) = P^T \left(\frac{\partial^2 f}{\partial z_i \partial z_j}\right) P,$$

where P is the change of coordinate matrix expressing the z_i in terms of the x_j.

In the 1-dimensional case, the determinant of the Hessian is simply the value of the second derivative of the function at p. Let us examine the functions f and g from Examples 1.2 and 1.3 above.

Example 1.6. For $f(x, y) = x^2 + \cos y$ the Hessian at an arbitrary point $(x, y) \in \mathbb{R}^2$ has determinant

$$H_f(x, y) = \det\begin{pmatrix} 2 & 0 \\ 0 & -\cos y \end{pmatrix} = -2\cos y.$$

At the origin, we have $H_f(0, 0) = -2$.

Example 1.7. For $g(x, y) = x^3 - 3xy^2$ the Hessian at $(x, y) \in \mathbb{R}^2$ has determinant

$$H_g(x, y) = \det\begin{pmatrix} 6x & -6y \\ -6y & -6x \end{pmatrix} = -36x^2 - 36y^2.$$

At the origin, we have $H_g(0, 0) = 0$.

Thus, we see that the Hessian plays the role for arbitrary smooth functions that the second derivative does for functions of a single variable. Moreover, the Hessian allows us to classify critical points in the following manner.

Definition 1.8. A critical point p of $f : M \to \mathbb{R}$ is *nondegenerate* if $H_f(p) \neq 0$. Since this condition is independent of the coordinate system near p (Remark 1.5), this notion is well-defined. Moreover, since coordinate changes are described by multiplication by a matrix and its transpose, the sign of $H_f(p)$ is independent of the coordinate system.

Definition 1.9. A smooth function $h : M \to \mathbb{R}$ is a *Morse function* if all of its critical points are nondegenerate.

It is not obvious that we can always construct Morse functions on a given manifold M. We will leave this question for a subsequent chapter. For now, let us look at some examples in detail.

In light of the calculations in Examples 1.6 and 1.7 above, we see that $f(x, y) = x^2 + \cos y$ is a Morse function on $M = \mathbb{R}^2$, while $g(x, y) = x^3 - 3xy^2$ is not. Throughout this text we shall mostly be interested in compact manifolds. Let us consider such an example.

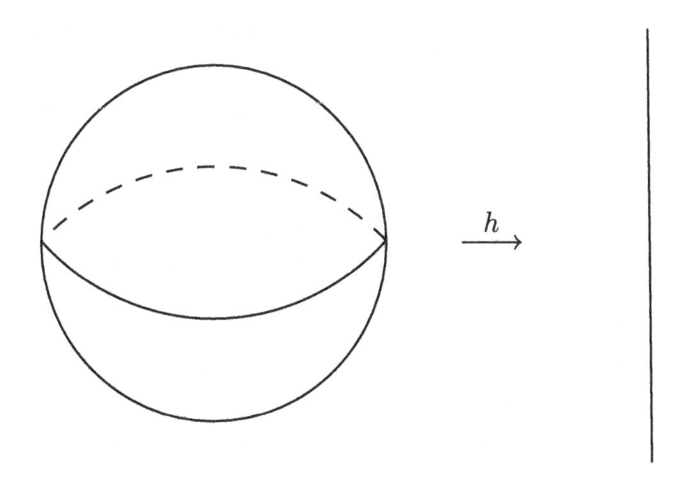

Fig. 1.3 The height function on the sphere

Example 1.10. Let S be the unit sphere in $\mathbb{R}^3$: $S = \{(x, y, z) \in \mathbb{R}^3 : x^2 + y^2 + z^2 = 1\}$. Define $h : S \to \mathbb{R}$ by $h(x, y, z) = z$. This is the *height function* on S. By examining Figure 1.3 it is intuitively obvious that h has exactly two critical points, namely the north and south poles. Let us perform the requisite calculations to show this.

Parametrize the upper hemisphere via the coordinate functions $(x, y, \sqrt{1-x^2-y^2})$, as (x, y) vary over the unit disk in $\mathbb{R}^2$. In terms of these coordinates, the function h takes the form $h(x, y) = \sqrt{1-x^2-y^2}$ and the north pole is at the origin of this coordinate system. We have the following partial derivatives:

$$\frac{\partial h}{\partial x} = \frac{-x}{\sqrt{1-x^2-y^2}} \quad \text{and} \quad \frac{\partial h}{\partial y} = \frac{-y}{\sqrt{1-x^2-y^2}}.$$

Thus, we see that the only critical point in this neighborhood is at the north pole, where $x = y = 0$. The Hessian of h has determinant

$$H_h(x, y) = \det \begin{pmatrix} \frac{y^2-1}{(1-x^2-y^2)^{3/2}} & \frac{-xy}{(1-x^2-y^2)^{3/2}} \\ \frac{-xy}{(1-x^2-y^2)^{3/2}} & \frac{x^2-1}{(1-x^2-y^2)^{3/2}} \end{pmatrix} = \frac{1}{(1-x^2-y^2)^2}.$$

Since $H_h(0, 0) \neq 0$, we see that the north pole is a nondegenerate critical point. Since $-h$ parametrizes the southern hemisphere, we may deduce that the south pole is a nondegenerate critical point as well.

Note that the Hessian is undefined along the equator since we have $x^2+y^2 = 1$ at these points. Strictly speaking, then, we must show that there are no critical points of h along the equator. This may be done by parametrizing the eastern and western hemispheres, and then noting that on these coordinate patches the function h takes the form $h(y, z) = \pm z$. As this function has no critical points, we are finished.

1.3 Morse functions on surfaces

Let us examine Morse functions on surfaces a bit further. Consider again the unit sphere S in $\mathbb{R}^3$. The height function h has the form $h(x, y) = \sqrt{1-x^2-y^2}$ with respect to the standard coordinates on the upper hemisphere. Consider the Taylor expansion of h in a neighborhood near $(0, 0)$:

$$h(x, y) = 1 - \frac{x^2}{2} - \frac{y^2}{2} + \text{h.o.t.}$$

where the higher order terms are all at least cubic in x and y. It follows that we may write

$$h(x, y) = 1 - x^2 p_1(x, y) - y^2 p_2(x, y)$$

for some smooth functions $p_1(x, y)$ and $p_2(x, y)$. These functions are not unique, but they are both positive in some neighborhood of the origin. Introduce new coordinates u and v:

$$u = x\sqrt{p_1(x, y)}$$
$$v = y\sqrt{p_2(x, y)}.$$

Then in terms of these new coordinates, we see that there is a neighborhood of the origin so that the height function may be expressed in local coordinates (u, v) as

$$h(u, v) = 1 - u^2 - v^2.$$

Similarly, the height function on the lower hemisphere may be taken as $-h(u, v) = -1 + u^2 + v^2$, at least near the origin. The graphs of these functions are paraboloids opening downward and upward, respectively, and in a small enough neighborhood of the poles they are indistinguishable from spherical caps. In Chapter 2 we shall prove the Morse Lemma, which tells us that we may always choose coordinates near a critical point so that a Morse function has only quadratic terms.

Before proceeding, we need a definition.

Definition 1.11. Suppose $f : M \to \mathbb{R}$ is a smooth function on an n-dimensional manifold M. If $a \in \mathbb{R}$, define the *sublevel set* M_a by

$$M_a = f^{-1}(-\infty, a] = \{x \in M : f(x) \leq a\}.$$

Sublevel sets are closed subsets of M. They are important objects of study in Morse theory. To see why, let us consider the sublevel sets of the height function h on the sphere S. When $a < -1$, the sublevel set S_a is empty. The set S_{-1} consists of only the south pole. For $-1 < a < 1$, the sublevel set S_a is a disc with center the south pole. From a topological point of view, this is not a particularly interesting space since it is contractible to a point. However, once we reach $a = 1$, something important happens: the sublevel set S_1 is the whole sphere and its topology is significantly different from the sublevel sets S_a for $a < 1$. Note that we have discrete jumps in the topology of the sublevel sets exactly when we pass critical values for our function h.

A more dramatic example of this is given by the torus. Figure 1.4 shows a torus T along with a height function $f : T \to \mathbb{R}$. Note that this is a Morse function.

The function f has critical points at p, q, r, and s. We may assume $f(p) = 0$. The sublevel set T_0 is simply the point p and for $a < f(q)$, we see that T_a is a disc centered at p. Once we reach $f(q)$, however, we have a much different scenario. If $f(q) < a < f(r)$, then the sublevel set T_a is a cylinder. Upon reaching $f(r)$, the topology changes again and for $f(r) < a < f(s)$, T_a is a torus with a disc removed. Topologically, this is a one-point union of two circles and is distinct from the sublevel sets in the interval before $f(r)$, which are topologically a single circle. Finally, passing $f(s)$ yields sublevel sets that are all of T. Figure 1.5 shows the possibilities.

Note that the critical points of f come in three varieties. At p, the surface looks like a paraboloid opening upward; that is, $f(u, v) = u^2 + v^2$ with respect to some local coordinates u, v. At q, the surface looks like a saddle and $f(u, v) = f(q) - u^2 + v^2$; a similar result holds at r. Note that in these cases there is a direction in which f increases and an orthogonal one in which f decreases near q

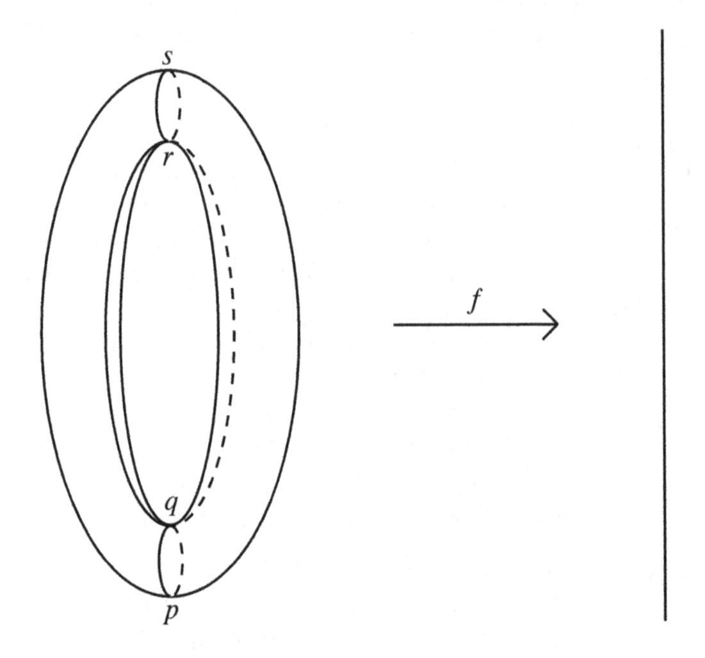

Fig. 1.4 The height function on the torus

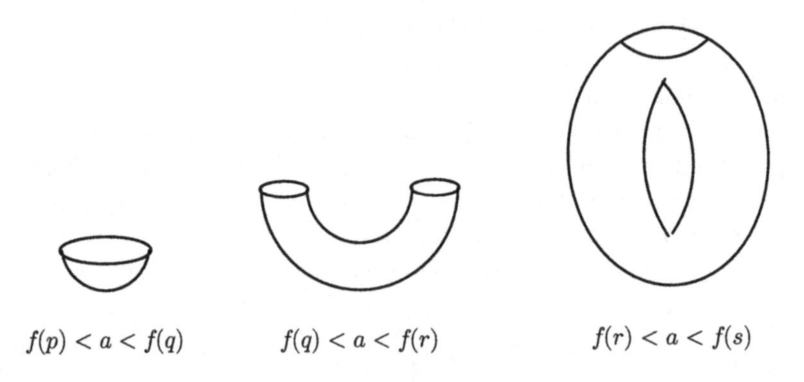

Fig. 1.5 The sublevel sets for the height function on the torus

and r. At the global maximum, s, the surface is a paraboloid opening downward and $f(u, v) = f(s) - u^2 - v^2$.

The torus with its height function is the essential example in Morse theory. In subsequent chapters we shall show that the critical points of a Morse function f may be described in an analogous fashion to the height function on the torus

(Morse Lemma) and that the topology of the manifold is determined by the types of critical points of f. The torus example also illustrates a phenomenon that holds in all dimensions: suppose there are no critical values of the function f between a and b. Then the sublevel sets M_a and M_b are homeomorphic. In fact, M_a is a deformation retract of M_b; that is, there is a map $d : M_b \to M_a$ such that d restricted to $M_a \subset M_b$ is the identity on M_a. It is truly remarkable that so much information may be deduced from the critical points of a smooth function.

1.4 Exercises

(1) View the torus as the product $S^1 \times S^1$ and use coordinates (θ, ϕ) to describe points. Let $R > r > 0$ be real numbers and define $f : S^1 \times S^1 \to \mathbb{R}$ by

$$f(\theta, \phi) = (R + r\cos\phi)\cos\theta.$$

Show that f is a Morse function, find all the critical points, and classify them by index.

(2) Consider the function $f : \mathbb{R}^2 \to \mathbb{R}^2$ defined by $f(x, y)$. Compute the determinant of the Hessian matrix at any point (x, y). Is the origin a nondegenerate critical point?

(3) Repeat the previous exercise with $f(x, y) = (x + y)^2$.

Chapter 2

Fundamental Results in Morse Theory

Now that we have examined some simple examples, we proceed with some general results. The first step is to characterize critical points of Morse functions.

2.1 The Morse Lemma

In Section 1.3 we examined the height function on the unit sphere in $\mathbb{R}^3$ and showed that there were local coordinates (u, v) near the north pole for which the function takes the form

$$f(u, v) = 1 - u^2 - v^2.$$

This representation depended on two facts:

(1) Smooth functions have Taylor expansions;
(2) There were no mixed quadratic monomials (e.g. xy) in the original formula for the function.

The first of these facts implies that, locally, Morse functions have no linear terms in their Taylor expansions. It will take some work to eliminate the mixed quadratic factors for an arbitrary Morse function, but it is possible to do so. Note, it is instructive to consider the fact that the function $g(x, y) = xy$ is a Morse function.

Theorem 2.1 (Morse Lemma). *Let p be a nondegenerate critical point of $f : M \to \mathbb{R}$. Then there is a local coordinate system $(X_1, X_2, \ldots, X_n)$, where p corresponds to the origin, such that the representation of f in these coordinates has the form*

$$f = f(p) - X_1^2 - X_2^2 - \cdots - X_i^2 + X_{i+1}^2 + \cdots + X_n^2.$$

Proof. Choose a local coordinate system $(x_1, x_2, \ldots, x_n)$ around p. Note that, by replacing f by $f - f(p)$, we may assume $f(p) = 0$. Since p is a critical point, all the partial derivatives vanish:

$$\frac{\partial f}{\partial x_j}(0, 0, \ldots, 0) = 0.$$

Therefore, using the Taylor expansion of f at the origin, we see that we may write

$$f(x_1, x_2, \ldots, x_n) = \sum_{i=1}^{n} x_i g_i(x_1, \ldots, x_n)$$

for some smooth functions g_i satisfying

$$\frac{\partial f}{\partial x_j}(0, \ldots, 0) = g_j(0, \ldots, 0).$$

Since p is a critical point, both sides of this equation vanish, and we may take Taylor expansions of each g_i to obtain n smooth functions h_{ij}, $1 \leq j \leq n$, such that

$$g_i(x_1, \ldots, x_n) = \sum_{j=1}^{n} x_j h_{ij}(x_1, \ldots, x_n)$$

in a neighborhood of the origin. We therefore may write

$$f(x_1, \ldots, x_n) = \sum_{i=1}^{n} \sum_{j=1}^{n} x_i x_j h_{ij}(x_1, \ldots, x_n).$$

Now, set $H_{ij} = (h_{ij} + h_{ji})/2$. Then f takes the form

$$f(x_1, \ldots, x_n) = \sum_{i=1}^{n} \sum_{j=1}^{n} x_i x_j H_{ij}(x_1, \ldots, x_n),$$

where we now have the symmetry relation $H_{ij} = H_{ji}$ for all i, j.

We may think of this as representing f by a quadratic form. Luckily, such forms are well understood. If we compute the second partial derivatives of f in the above form we see that

$$\frac{\partial^2 f}{\partial x_i \partial x_j}(0, \ldots, 0) = 2H_{ij}(0, \ldots, 0).$$

Since the critical point p is nondegenerate, the matrix $(H_{ij}(0, \ldots, 0))$ is nonsingular. Moreover, we may assume that

$$\frac{\partial^2 f}{\partial x_1^2}(0, \ldots, 0) \neq 0.$$

Indeed, even if all the second partials $\partial^2 f/\partial x_j^2$ vanish at p, there is some $H_{k\ell}$ that is nonzero in a neighborhood of $(0, \ldots, 0)$ and the linear change of coordinates

$$Y_k = x_k - x_\ell \qquad Y_\ell = x_k + x_\ell$$

yields $\partial^2 f/\partial Y_k^2 \neq 0$ at the origin. Permuting these new coordinates allows us to proceed with our assumption.

Now, since $\partial^2 f/\partial x_1^2(0, \ldots, 0) \neq 0$, we see that H_{11} does not vanish in a neighborhood of the origin (by continuity). Introduce a new coordinate system by leaving $x_2, \ldots, x_n$ unchanged and setting

$$X_1 = \sqrt{|H_{11}|}\left(x_1 + \sum_{i=1}^{n} x_i \frac{H_{1i}}{H_{11}}\right).$$

The Jacobian of the transformation from our old coordinates to this new set has nonzero determinant, so $(X_1, x_2, \ldots, x_n)$ is indeed a local coordinate system. Now observe that

$$X_1^2 = \begin{cases} H_{11}x_1^2 + 2\sum_{i=2}^{n} x_1 x_i H_{1i} + (\sum_{i=2}^{m} x_i H_{1i})^2/H_{11} & H_{11} > 0, \\ -H_{11}x_1^2 - 2\sum_{i=2}^{n} x_1 x_i H_{1i} - (\sum_{i=2}^{m} x_i H_{1i})^2/H_{11} & H_{11} < 0. \end{cases}$$

Using this we may write f in terms of the coordinates $(X_1, x_2, \ldots, x_n)$ to get

$$f = \begin{cases} X_1^2 + \sum_{i=2}^{n} \sum_{j=2}^{n} x_i x_j H_{ij} - (\sum_{i=2}^{n} x_i H_{1i})^2/H_{11} & H_{11} > 0, \\ -X_1^2 + \sum_{i=2}^{n} \sum_{j=2}^{n} x_i x_j H_{ij} - (\sum_{i=2}^{n} x_i H_{1i})^2/H_{11} & H_{11} < 0. \end{cases}$$

Note that after the X_1^2 term, the terms only involve $x_2, \ldots, x_n$. We may then inductively proceed to put f into the required form. □

Here are two immediate consequences of the Morse Lemma.

Corollary 2.2. *The critical points of a Morse function are isolated.*

Proof. Let $f : M \to \mathbb{R}$ be a Morse function and suppose p is a critical point of f. Then in some neighborhood of p, f has the form $f = f(p) - x_1^2 - \cdots - x_i^2 + x_{i+1}^2 + \cdots + x_n^2$. But there are no other critical points in this neighborhood, as a direct calculation shows. □

Corollary 2.3. *A Morse function f on a compact manifold M has only finitely many critical points.*

Proof. Suppose there are infinitely many critical points of f:

$$p_1, p_2, p_3, \ldots$$

Since M is compact, this sequence has a convergent subsequence

$$p_{i_1}, p_{i_2}, p_{i_3}, \ldots,$$

which we may assume lies in a neighborhood U of the limit point p. In this neighborhood, we may consider the partial derivatives $\dfrac{\partial f}{\partial x_i}$. These vanish at the p_{i_k} and since they are continuous functions, they also vanish at the limit point p. But then p is a critical point of f which has a neighborhood U containing infinitely many critical points of f, contrary to the fact that the critical points of f are isolated. □

The number i of minus signs in the above representation is the number of negative diagonal entries in the (symmetric) matrix of second partial derivatives after diagonalization. Sylvester's Law [Lang (2004)] implies that i does not depend on how this matrix is diagonalized; that is, i is determined by f and the critical point p.

Definition 2.4. The number i of minus signs in the representation of the function is called the *index* of the critical point. It is an integer between 0 and n, inclusive.

Example 2.5. *The n-sphere.* Consider the sphere $S^n = \{(x_1, \ldots, x_{n+1}) \in \mathbb{R}^{n+1} : x_1^2 + \cdots + x_{n+1}^2 = 1\}$. The height function

$$h : S^n \to \mathbb{R}, \qquad f(x_1, \ldots, x_n, x_{n+1}) = x_{n+1}$$

is obviously a Morse function with critical points the north and south poles: $p = (0, \ldots, 0, 1)$ and $q = (0, \ldots, 0, -1)$. As in the case $n = 2$, it is easy to see that we may choose local coordinates around p and q so that h has the form

$$h(u_1, \ldots, u_n) = \begin{cases} 1 - u_1^2 - \cdots - u_n^2 & \text{near } p \\ 1 + u_1^2 + \cdots + u_n^2 & \text{near } q. \end{cases}$$

It follows that the index of p is n and the index of q is 0.

Note that index 0 critical points must be local minima of the Morse function and that index n critical points ($n = \dim M$) are local maxima. We have just seen that the n-sphere has a Morse function with exactly two critical points. We now prove a converse of this statement.

Theorem 2.6. *Suppose M is a compact n-manifold and that $f : M \to \mathbb{R}$ is a Morse function with exactly two critical points. Then M is homeomorphic to the sphere S^n.*

Proof. Since M is compact and f is continuous, there are points p and q where f attains its maximum and minimum values, respectively. These must be critical points of f as well. Choose local coordinates $(u_1, \ldots, u_n)$ in a neighborhood of p such that $f = f(p) - u_1^2 - \cdots - u_n^2$ and coordinates $(v_1, \ldots, v_n)$ in a neighborhood of q such that $f = f(q) + v_1^2 + \cdots + v_n^2$. Let U_ε be the subset of M consisting of points x satisfying

$$f(p) - \varepsilon \leq f(x) \leq f(p)$$

and let V_ε be the set of y satisfying

$$f(q) \leq f(y) \leq f(q) + \varepsilon.$$

Now, U_ε and V_ε are both homeomorphic to the closed disk D^n with boundary sphere S^{n-1}. If we choose ε small enough we may assume $U_\varepsilon \cap V_\varepsilon = \emptyset$. Let M_0 be the subset of M obtained by removing the interiors of U_ε and V_ε. Then M_0 is a compact manifold with boundary homeomorphic to two disjoint copies of S^{n-1}. Note that f takes on constant values on each boundary component and has no critical points in the interior of M_0. We will show later (Theorem 2.17) that this implies that M_0 is homeomorphic to the direct product of one of the boundary spheres (say ∂V_ε) with the unit interval. Now construct a space N_0 by attaching V_ε to M_0 along the common boundary they share. This is homeomorphic to the disc D^n since it fills in the interior hole of the annulus M_0. We then construct M by pasting U_ε to N_0 along their common boundary. This shows that M is homeomorphic to the sphere S^n. □

Remark 2.7. More is true, at least when $n \leq 6$. In those cases, we can actually say that M is *diffeomorphic* to S^n. Trouble arises once $n = 7$, however, as Milnor showed by constructing many distinct differentiable structures on S^7 [Milnor (1956)].

2.2 Existence of Morse functions

Up until now we have considered explicit examples of Morse functions on well-known manifolds. This begs the question, however, of whether or not an arbitrary manifold supports a Morse function. Smooth functions exist on an arbitrary manifold (constant functions, for example), but we need to prove that we can always find a Morse function. In this section, we will actually prove more: given any smooth function f on a compact manifold M, there is a Morse function g arbitrarily close to f in a sense we will make explicit.

Consider the unit circle S^1 in the plane with the constant function $f = 0$. Let $\varepsilon > 0$ and consider the coordinate u on the upper semicircle, $-1/\varepsilon < u < 1/\varepsilon$. The function g on the circle defined by

$$g(u) = \varepsilon(1 - \varepsilon^2 u^2)$$

is a Morse function, as can be seen through a brief calculation. Indeed, it is the square of the height function of the ellipse $\varepsilon^2 x^2 + y^2/\varepsilon^2 = 1$. Observe the following inequalities for all $p \in S^1$:

$$\begin{aligned} |f(p) - g(p)| &= \varepsilon(1 - \varepsilon^2 u^2) < \varepsilon \\ \left|\tfrac{df}{du}(p) - \tfrac{dg}{du}(p)\right| &= \varepsilon(2u\varepsilon^2) \quad < \varepsilon \\ \left|\tfrac{d^2 f}{du^2}(p) - \tfrac{d^2 g}{du^2}(p)\right| &= 2\varepsilon^3 \quad < \varepsilon \end{aligned}$$

where the last two inequalities hold for ε sufficiently small. The reader should imagine that we have reparametrized the circle so that it looks like a narrow ellipse whose y-coordinates lie in the band $-\varepsilon \le y \le \varepsilon$, and then the height function on this ellipse is within ε of the constant function f. Here is the formal definition we will use for "closeness" of two functions.

Definition 2.8. Suppose f and g are smooth functions on a compact set K contained in a coordinate neighborhood of M. If $\varepsilon > 0$, we say that f is a (C^2, ε)-approximation of g in K if the following inequalities hold for every $p \in K$:

$$\begin{aligned} |f(p) - g(p)| &< \varepsilon \\ \left|\frac{\partial f}{\partial x_i}(p) - \frac{\partial g}{\partial x_i}(p)\right| &< \varepsilon \qquad i = 1, \ldots, n \\ \left|\frac{\partial^2 f}{\partial x_i \partial x_j}(p) - \frac{\partial^2 g}{\partial x_i \partial x_j}(p)\right| &< \varepsilon \qquad i, j = 1, \ldots, n. \end{aligned}$$

We are now ready to state the main theorem of this section.

Theorem 2.9. *Suppose g is a smooth function on a compact manifold M. Then there is a Morse function f on M such that f is a (C^2, ε)-approximation of g on M for sufficiently small $\varepsilon > 0$.*

We need some preliminary results before proving Theorem 2.9.

Lemma 2.10. *Let U be an open set in $\mathbb{R}^m$ and suppose $f : U \to \mathbb{R}$ is a smooth function. Then for some real numbers $a_1, a_2, \ldots, a_m$, the function*

$$g(x_1, \ldots, x_m) = f(x_1, \ldots, x_m) - (a_1x_1 + a_2x_2 + \cdots + a_mx_m)$$

is a Morse function on U. Moreover, we may choose the a_i to be arbitrarily small in absolute value.

Proof. First recall Sard's Theorem [Milnor (1965a)]: if $r : U \to \mathbb{R}^m$ is a smooth function, then the set of critical values has measure zero in $\mathbb{R}^m$ (a *critical value* is the value of r at a critical point; see Appendix A for further details). This means that we may always find points in any open set in $\mathbb{R}^m$ which are *not* critical values for r. Given our smooth function f, define a map $h : U \to \mathbb{R}^m$ by

$$h = \begin{pmatrix} \frac{\partial f}{\partial x_1} \\ \frac{\partial f}{\partial x_2} \\ \vdots \\ \frac{\partial f}{\partial x_m} \end{pmatrix}.$$

The Jacobian matrix of h at a point p_0 is

$$J_h(p_0) = \begin{pmatrix} \frac{\partial^2 f}{\partial x_1^2}(p_0) & \cdots & \frac{\partial^2 f}{\partial x_1 \partial x_m}(p_0) \\ & \ddots & \\ \vdots & \frac{\partial^2 f}{\partial x_i \partial x_j}(p_0) & \vdots \\ & \ddots & \\ \frac{\partial^2 f}{\partial x_m \partial x_1}(p_0) & \cdots & \frac{\partial^2 f}{\partial x_m^2}(p_0) \end{pmatrix}.$$

Observe that this is equal to the Hessian of f and so p_0 is a critical point of $h : U \to \mathbb{R}$ if and only if $\det H_f(p_0) = 0$.

By Sard's Theorem, we may choose a point $(a_1, a_2, \ldots, a_m) \in \mathbb{R}^m$ which is not a critical value of h; moreover, we may choose the a_i to be arbitrarily small in absolute value. We claim that with this choice of a_i, the function g defined above is a Morse function on U. To see this note that if p_0 is a critical point of g, then since

$$0 = \frac{\partial g}{\partial x_i}(p_0) = \frac{\partial f}{\partial x_i}(p_0) - a_i$$

we have

$$h(p_0) = \begin{pmatrix} a_1 \\ a_2 \\ \vdots \\ a_m \end{pmatrix}.$$

But this point is not a critical value of $h : U \to \mathbb{R}^m$ and so p_0 is not a critical point of h. Therefore, $\det H_f(p_0) \neq 0$. Since f and g differ only by a linear function, their Hessians agree, and then we have

$$\det H_f(p_0) = \det H_g(p_0).$$

It follows that p_0 is a nondegenerate critical point of g, and so g is a Morse function on U.

$\square$

Before proving Theorem 2.9 we need to clarify what we mean by a (C^2, ε)-approximation on a compact manifold M (rather than on a compact subset K). We can cover M by an infinite number of coordinate neighborhoods (e.g., by δ-balls around each point M). Given such a cover, we may choose a finite subcover $U_1, U_2, \ldots, U_k$. We may then choose compact sets $K_i \subset U_i$ such that

$$M = K_1 \cup K_2 \cup \cdots \cup K_k.$$

Definition 2.11. A function $f : M \to \mathbb{R}$ is a (C^2, ε)-approximation of a function $g : M \to \mathbb{R}$ if f is such an approximation on each K_i.

Lemma 2.12. *Let C be a compact set in M. Suppose that $g : M \to \mathbb{R}$ has no degenerate critical point in C. Then there is a sufficiently small $\varepsilon > 0$ such that any (C^2, ε)-approximation f of g has no degenerate critical point in C.*

Proof. Choose a coordinate neighborhood U_i in the finite cover of M. Let $(x_1, \ldots, x_m)$ be the coordinate system in U_i. It is easy to show that there are no degenerate critical points of g in $C \cap K_i$ if and only if the following condition holds in $C \cap K_i$:

$$\left|\frac{\partial g}{\partial x_1}\right| + \cdots + \left|\frac{\partial g}{\partial x_m}\right| + \left|\det\left(\frac{\partial^2 g}{\partial x_i \partial x_j}\right)\right| > 0.$$

For a (C^2, ε)-approximation f of g, the similar inequality

$$\left|\frac{\partial f}{\partial x_1}\right| + \cdots + \left|\frac{\partial f}{\partial x_m}\right| + \left|\det\left(\frac{\partial^2 f}{\partial x_i \partial x_j}\right)\right| > 0$$

holds for all ε small enough (this is by the definition of (C^2, ε)-approximation). Thus, f has no degenerate critical point in $C \cap K_i$ and therefore f has no degenerate critical point in

$$C = \bigcup_{i=1}^{k} (C \cap K_i).$$

$\square$

We may now prove Theorem 2.9. Cover M by coordinate neighborhoods U_i with compact subsets K_i also providing a cover. We proceed inductively to construct functions f_i from g on M with no degenerate critical points in $K_1 \cup \cdots \cup K_i$. Since $M = K_1 \cup \cdots \cup K_k$, this will complete the proof. Set f_0 to be g and denote by C_i the set $K_1 \cup \cdots \cup K_i$. This is a compact set. Also, set $C_0 = \emptyset$.

Suppose we have constructed $f_{i-1} : M \to \mathbb{R}$ with no degenerate critical points in C_{i-1}. Consider the coordinate neighborhood U_i; denote the coordinates by $(x_1, \ldots, x_m)$. By Lemma 2.10 there are real numbers $a_1, \ldots, a_m$ with small enough absolute value so that

$$g_{i-1}(x_1, \ldots, x_m) = f_{i-1}(x_1, \ldots, x_m) - (a_1x_1 + \cdots + a_mx_m)$$

is a Morse function on U_i. Let us modify g_{i-1} so that the linear piece makes sense outside U_i. We do this via an auxiliary function $h_i : U_i \to \mathbb{R}$ which has the following properties:

(1) $0 \leq h_i \leq 1$;
(2) h_i takes the value 1 on some open neighborhood V_i of K_i;
(3) h_i takes the value 0 outside some compact set L_i containing V_i.

That such a function h exists is a standard result in manifold theory (see, e.g., [Munkres (1963)]). Construct a function f_i on M as follows:

$$f_i = \begin{cases} f_{i-1}(x_1, \ldots, x_m) - (a_1x_1 + \cdots + a_mx_m)h_i(x_1, \ldots, x_m) & \text{in } U_i \\ f_{i-1}(x_1, \ldots, x_m) & \text{outside } L_i. \end{cases}$$

Note that f_i agrees with the function g_{i-1} in some neighborhood of the compact set K_i. Thus, f_i is a Morse function on K_i.

We must show that f_i is a (C^2, ε)-approximation of f_{i-1}. First note that on U_i we have for $j, k = 1, \ldots, m$

$$|f_{i-1}(p) - f_i(p)| = |(a_1x_1 + \cdots + a_mx_m)|h_i(p)$$

$$\left|\frac{\partial f_{i-1}}{\partial x_j}(p) - \frac{\partial f_i}{\partial x_j}(p)\right| = \left|a_j h_i(p) + (a_1x_1 + \cdots + a_mx_m)\frac{\partial h_i}{\partial x_j}(p)\right|$$

$$\left|\frac{\partial^2 f_{i-1}}{\partial x_j \partial x_k}(p) - \frac{\partial^2 f_i}{\partial x_j \partial x_k}(p)\right| = \left|a_j \frac{\partial h_i}{\partial x_k}(p) + a_k \frac{\partial h_i}{\partial x_j}(p) + (a_1x_1 + \cdots + a_mx_m)\frac{\partial^2 h_i}{\partial x_j \partial x_k}(p)\right|.$$

Since the function h_i satisfies $0 \leq h_i \leq 1$ and vanishes outside the compact set L_i, the absolute values of its first and second derivatives are bounded. We may therefore make the right-hand sides of the above equations arbitrarily small by choosing the a_j small enough. This implies that f_i can be made (C^2, ε)-close to f_{i-1} in K_i.

Now, if K_j is another such set, let $(y_1, \ldots, y_m)$ be the coordinates in U_j. We must compute the differences between the first and second derivatives of f_i and f_{i-1}

in K_j. First note that $f_i = f_{i-1}$ outside the set L_i so that we need only estimate the difference on $K_j \cap L_i$, which is contained in the intersection $U_j \cap U_i$. Note that the difference on $U_j \cap U_i$ may be expressed by the equations on the right side above with a suitable change of coordinates between $(x_1, \ldots, x_m)$ and $(y_1, \ldots, y_m)$. Since we are working on the compact set $K_j \cap L_i$, the absolute value of each component of the corresponding Jacobian transformation cannot exceed a fixed universal bound. Thus, if we make the a_ℓ small enough, we can again make the corresponding set of quantities as small as we wish on $K_j \cap L_i$. Since f_i and f_{i-1} agree outside L_i, we see that f_i is (C^2, ε)-close to f_{i-1} provided the a_ℓ are sufficiently small. Repeating this process for each $j = 1, 2, \ldots, k$, we see that f_i is a (C^2, ε)-approximation of f_{i-1} for any positive ε.

By the inductive hypothesis, f_{i-1} does not have degenerate critical points in $C_{i-1} = K_1 \cup \cdots \cup K_{i-1}$. By Lemma 2.12, f_i does not have degenerate critical points in C_{i-1}. But we also constructed f_i to have no degenerate critical points in K_i and so it has no such points in C_i. This completes the inductive step.

Proceeding inductively, we therefore construct a Morse function f_k on M, which can be made (C^2, ε)-close to g for any specified ε by making the correct choices in each inductive step. This completes the proof of Theorem 2.9. □

2.3 Gradient-like vector fields

A *vector field* on M is an assignment of a tangent vector $\mathbf{v}$ to each point $p \in M$. If U is a coordinate neighborhood with coordinate system $(x_1, \ldots, x_m)$, then a vector field X on U may be written

$$X = \chi_1 \frac{\partial}{\partial x_1} + \chi_2 \frac{\partial}{\partial x_2} + \cdots + \chi_m \frac{\partial}{\partial x_m},$$

where the χ_i are functions defined in U. The vector field X is *smooth* if each of the functions χ_i is smooth in U.

Example 2.13. If f is a smooth function in U with coordinates $(x_1, x_2, \ldots, x_m)$, we define a vector field X_f in U by

$$X_f = \frac{\partial f}{\partial x_1}\frac{\partial}{\partial x_1} + \frac{\partial f}{\partial x_2}\frac{\partial}{\partial x_2} + \frac{\partial f}{\partial x_m}\frac{\partial}{\partial x_m}.$$

The vector field X_f is called the *gradient vector field* of f.

Just as we can differentiate a function in the direction of a tangent vector at a point (see Appendix A), we may differentiate a function in the direction of a vector field. In the case of the gradient field X_f, we have

$$\begin{aligned} X_f \cdot f &= \left(\sum_{i=1}^{m} \frac{\partial f}{\partial x_i} \frac{\partial}{\partial x_i} \right) \cdot f \\ &= \sum_{i=1}^{m} \left(\frac{\partial f}{\partial x_i} \right)^2 \\ &\geq 0. \end{aligned}$$

Observe that $(X_f \cdot f)(p) > 0$ unless p is a critical point of f. Thus we see that the gradient vector field of f points in the direction in which f is increasing. At this stage it is instructive to consider the gradient vector field of a Morse function

$$f = -x_1^2 - \cdots - x_i^2 + x_{i+1}^2 + \cdots x_m^2$$

defined in a coordinate neighborhood of a critical point p. The gradient of f is

$$-2x_1 \frac{\partial}{\partial x_1} - \cdots - 2x_i \frac{\partial}{\partial x_i} + 2x_{i+1} \frac{\partial}{\partial x_{i+1}} + \cdots + 2x_m \frac{\partial}{\partial x_m}.$$

The vector fields for various values of i are shown in Figure 2.1.

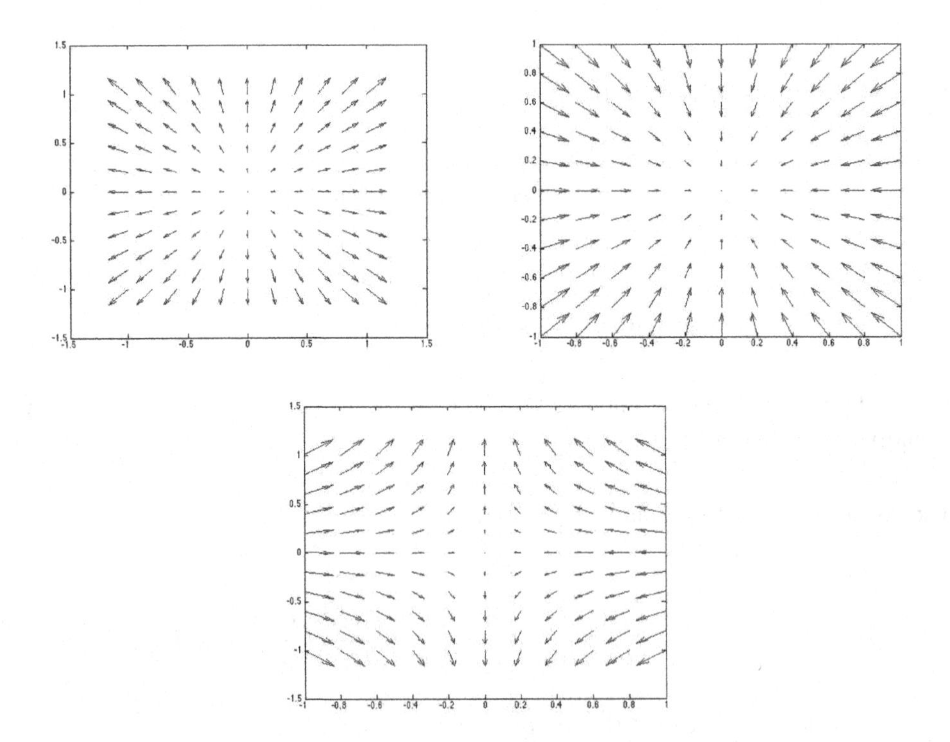

Fig. 2.1 The gradient vector fields near a critical point of index 0 (top left), index m (top right), and index i (bottom). In the index i case, one should think of the first i coordinates as lying along the horizontal axis and the last $m - i$ along the vertical.

Definition 2.14. Let $f : M \to \mathbb{R}$ be a Morse function on M and suppose X is a vector field on M. Then X is a *gradient-like vector field for* f if

(1) $X \cdot f > 0$ away from the critical points of f; and
(2) If p is a critical point of f of index i, then p has a sufficiently small coordinate neighborhood U such that f has a standard form

$$f = f(p) - x_1^2 - \cdots - x_i^2 + x_{i+1}^2 + \cdots + x_m^2$$

and X can be written as its gradient field

$$X = -2x_1 \frac{\partial}{\partial x_1} - \cdots - 2x_i \frac{\partial}{\partial x_i} + 2x_{i+1} \frac{\partial}{\partial x_{i+1}} + \cdots + 2x_m \frac{\partial}{\partial x_m}.$$

Gradient-like vector fields always point in the direction in which f is increasing. As a simple example, consider the height function h on the sphere S^2. The function h increases most rapidly along longitudinal great circles. A gradient-like field for h need not consist of velocity vectors for these curves, but it is required to point at least slightly toward the north pole at each point. If any tangent vector pointed orthogonal to a longitude or toward the south pole, then it would fail criterion (i).

Theorem 2.15. *If $f : M \to \mathbb{R}$ is a Morse function on a compact manifold M, then there exists a gradient-like vector field for f.*

Proof. The proof is much like the proof of Theorem 2.9; we therefore only provide a sketch. We may cover M by a finite number of coordinate neighborhoods $U_1, U_2, \ldots, U_\ell$ and choose a compact set K_i in each U_i so that the collection of K_i cover M. Moreover, we may assume that each critical point p has a neighborhood contained in exactly one U_j and that f has a standard form in U_j. As in the proof of Theorem 2.9, we have functions $h_j : U_j \to \mathbb{R}$ which satisfy $0 \leq h_j \leq 1$, $h_j = 1$ in some neighborhood V_j of K_j and $h_j = 0$ outside a compact set L_j containing V_j and contained in U_j. We extend h_j to all of M by setting it to be 0 outside U_j. Now, in each U_j, define X_j to be the gradient of f. This satisfies $X_j \cdot f > 0$ away from the critical points. Define a vector field X on M by

$$X = \sum_{j=1}^{\ell} h_j X_j.$$

It is easy to check that $X \cdot f > 0$ away from the critical points of f.

At a critical point p, there is a small neighborhood V contained in only one of the U_i, and in this neighborhood, $X = h_i X_i = X_i$. It follows that X has the correct form in V and so is a gradient-like vector field for f. □

2.4 Integral curves

In Chapter 3, we will use the results of this section to obtain information about the topology of a manifold M from the critical points of a Morse function on M. A crucial concept is that of an *integral curve* on M.

Definition 2.16. A curve $\sigma : I \to M$ is an *integral curve* of a vector field X if

$$\frac{d\sigma}{dt}(t) = X_{\sigma(t)}$$

for all $t \in I$.

An integral curve is a flow line of a particle moving in M with X as its velocity vectors. Such lines exist: if M is a compact manifold without boundary, then there exists an integral curve $\sigma_p(t)$ for $-\infty < t < \infty$ with $\sigma_p(0) = p$ [Munkres (1963)].

Now suppose X is a gradient-like vector field for a Morse function f. If p is not a critical point of f, then the integral curve $\sigma_p(t)$ approaches critical points as $t \to \pm\infty$. Note that the curve never reaches the critical points since the vectors of X get smaller in magnitude.

If $[a, b]$ is a real interval, set

$$M_{[a,b]} = \{p \in M : a \leq f(p) \leq b\}.$$

Theorem 2.17. *If f has no critical value in $[a, b]$, then $M_{[a,b]}$ is diffeomorphic to the product*

$$f^{-1}(a) \times [0, 1].$$

Proof. Let X be a gradient-like vector field for f. Let C be the finite set of critical points of f and define a vector field Y on the open set $M - C$ by

$$Y = \frac{1}{X \cdot f} X.$$

Note that Y is defined on $M_{[a,b]}$. Let p be a point in $f^{-1}(a)$ and let $\sigma_p(t)$ be an integral curve of Y beginning at p. Then we have

$$\begin{aligned} \frac{d}{dt} f(\sigma_p(t)) &= \frac{d\sigma_p}{dt}(t) \cdot f \\ &= Y_{\sigma_p(t)} \cdot f \\ &= \frac{1}{X \cdot f} X \cdot f \\ &= 1. \end{aligned}$$

So we see that $\sigma_p(t)$ climbs upward at unit speed with respect to the function f (which we may think of as a height function in a certain direction). The curve $\sigma_p(t)$ reaches height $f = b$ at time $t = b - a$. Define $h : f^{-1}(a) \times [0, b - a] \to M_{[a,b]}$ by

$$h(p, t) = \sigma_p(t).$$

Since $\sigma_p(t)$ depends smoothly on p and t and since two distinct integral curves do not meet, it follows that h is a diffeomorphism. Since $[0, b - a]$ is diffeomorphic to $[0, 1]$, this completes the proof. □

2.5 Exercises

(1) Let $f : M \to \mathbb{R}$ be a Morse function on M with critical points $p_1, p_2, \ldots, p_k$. Prove that there is a Morse function g with the same critical points satisfying $g(p_i) \neq g(p_j)$ for $i \neq j$. Show also that g may be taken as close to f as we wish.

(2) Recall that the funciton $f : S^1 \times S^1 \to \mathbb{R}$ defined by

$$f(\theta, \phi) = (R + r\cos\phi)\cos\theta$$

is a Morse function. Compute ∇f. What are the integral curves of ∇f?

(3) Show that the map $g : S^1 \to \mathbb{R}$ defined by $g_+(u) = \varepsilon(1 - \varepsilon^2 u^2)$ on the upper semicircle $-1/\varepsilon < u < 1/\varepsilon$, and by $-g_+$ on the lower semicircle is a Morse function. Show that g is a (C^2, ε)-approximation to the constant function $f = 0$ on S^1.

Bibliographic notes

The results in this section are standard and may be found in any book on Morse theory (e.g., [Milnor (1963)], [Milnor (1965b)], [Matsumoto (1997)]) or even online. This is especially true of the Morse Lemma. We therefore did not provide explicit references. The proof of Theorem 2.9 follows the presentation in [Matsumoto (1997)] as there is little the author could do to improve it.

Chapter 3

Topological Consequences

In Chapter 1 we gave examples of Morse functions on surfaces and noted how the topology of the sublevel sets changed as we passed the function's critical points. In this chapter we shall investigate this phenomenon in general and prove powerful decomposition results for arbitrary smooth compact manifolds.

3.1 Handle decompositions of manifolds

Recall that if $f : M \to \mathbb{R}$ is a Morse function, the sublevel set M_a is defined as $M_a = \{p \in M : f(p) \leq a\}$. The first fundamental result about sublevel sets is the following theorem.

Theorem 3.1. *Suppose $a < b$ and that f has no critical values in the interval $[a, b]$. Then the sublevel sets M_a and M_b are diffeomorphic. In fact, more is true: M_a is a deformation retract of M_b.*

Proof. Let X be a gradient-like vector field for f. In the proof of Theorem 2.17 we used the vector field $Y = \dfrac{1}{X \cdot f} X$ to show that $M_{[a,b]}$ is diffeomorphic to the product $f^{-1}(a) \times [0, 1]$; denote this diffeomorphism by h. Recall that this involved using the flow lines of Y beginning at points of $f^{-1}(a)$. Clearly, we have $M_b = M_a \cup M_{[a,b]}$, where M_a and $M_{[a,b]}$ are glued along their common boundary $f^{-1}(a)$. We may therefore define a diffeomorphism $g : M_a \to M_b$ by taking g to be the identity on M_a and h on $M_{[a,b]}$. Since g and h agree on $f^{-1}(a)$ this map is well-defined and gives the required map. Also, note that $M_{[a,b]}$ retracts onto $f^{-1}(a)$; simply flow backwards along the flow lines of Y. It follows that M_a is in fact a deformation retract of M_b. □

Remark 3.2. Note that the proof of Theorem 3.1 did not use the fact that f is a Morse function. All that is required is the existence of a gradient-like vector field. It follows that Theorem 3.1 is valid for any smooth function, Morse or not: if $g : M \to \mathbb{R}$ is a smooth function, $[a, b]$ contains no critical values of f, and if

$f^{-1}[a,b]$ is compact, then M_a is a deformation retract of M_b, and the two spaces M_a and M_b are diffeomorphic.

Thus, we see that the only potentially interesting changes in the sublevel sets occur at critical values for f. Suppose that the critical points for f are $p_0, p_1, \ldots, p_k$ with corresponding critical values $c_i = f(p_i)$. By Exercise (1) in Chapter 2, we may assume that $c_i \neq c_j$ for $i \neq j$. Moreover, we assume we have ordered the critical points so that

$$c_0 < c_1 < \cdots < c_k.$$

Observe that $M_t = \emptyset$ for $t < c_0$ and that $M_t = M$ for $t > c_k$.

Consider first the case of the minimum value c_0. There are local coordinates about p_0 where f has the standard form

$$f = x_1^2 + x_2^2 + \cdots + x_m^2 + c_0.$$

If $\varepsilon > 0$ is small (in fact, any $\varepsilon < c_1 - c_0$ will work), then $M_{c_0-\varepsilon} = \emptyset$ and

$$M_{c_0+\varepsilon} = \{(x_1, \ldots, x_m) : x_1^2 + x_2^2 + \cdots + x_m^2 \leq \varepsilon\}.$$

That is, $M_{c_0+\varepsilon}$ is diffeomorphic to the m-disc D^m and f takes the minimum value c_0 at the origin. In the case of a surface ($m = 2$), we may visualize this set $M_{c_0+\varepsilon}$ as a bowl opening upward; the higher-dimensional case is analogous (but impossible to see).

Now, if the parameter t passes an index 0 critical value c_i, the same thing happens: we add an m-disc pointing upward and $M_{c_i+\varepsilon}$ is diffeomorphic to the disjoint union $M_{c_i-\varepsilon} \cup D^m$. We call such a disc, corresponding to a critical point of index 0, a 0-*handle.*

Consider now the case of the global maximum c_k. In local coordinates near p_k, f has the form

$$f = c_k - x_1^2 - x_2^2 - \cdots - x_m^2.$$

The index of p_k is therefore m. When $t > c_k$, $M_t = M$. For small ε, the sublevel set $M_{c_k-\varepsilon}$ is a manifold with boundary the set of points in M with $f(z) = c_k - \varepsilon$. The complement $M - M_{c_k-\varepsilon}$ is the set of points $(x_1, \ldots, x_m)$ with

$$x_1^2 + x_2^2 + \cdots + x_m^2 < \varepsilon;$$

that is, M is the union of $M_{c_k-\varepsilon}$ and an m-disc glued together along the boundary $(m-1)$-sphere $f^{-1}(c_k - \varepsilon)$. We call this disc an m-*handle*. Every time we pass a local maximum (index m critical point), the same thing happens: we add an m-disc pointing "downward" which caps off a component of the boundary of $M_{c_j-\varepsilon}$.

What about critical points of index i for $0 < i < m$? Choose local coordinates around such a point p_j and write f in standard form:

$$f = c_j - x_1^2 - \cdots - x_i^2 + x_{i+1}^2 + \cdots + x_m^2.$$

Choose a small value ε and a positive δ with $\delta \ll \varepsilon$. We assume the closed set $f^{-1}[c_j - \varepsilon, c_j + \varepsilon]$ contains no critical points except for p_j. We wish to understand how $M_{c_j+\varepsilon}$ is built from $M_{c_j-\varepsilon}$. The latter is obtained by setting $f \leq c_j - \varepsilon$ which in turn means

$$x_1^2 + \cdots + x_i^2 - x_{i+1}^2 - \cdots - x_m^2 \geq \varepsilon.$$

Consider the set of points satisfying

$$x_1^2 + \cdots + x_i^2 - x_{i+1}^2 - \cdots - x_m^2 \leq \varepsilon$$
$$x_{i+1}^2 + \cdots + x_m^2 \leq \delta.$$

We call this an *m-dimensional handle of index i*; it is diffeomorphic to the direct product $D^i \times D^{m-i}$. The i-disc

$$D^i \times 0 = \{(x_1, \ldots, x_i, 0, \ldots, 0) : x_1^2 + \cdots x_i^2 \leq \varepsilon\}$$

is called the *core* of the handle and the $(m-i)$-disc

$$0 \times D^{m-i} = \{(0, \ldots, 0, x_{i+1}, \ldots, x_m) : x_{i+1}^2 + \cdots x_m^2 \leq \delta\}$$

is called the *co-core*. Attach an i-handle to $M_{c_j-\varepsilon}$ (see Figure 3.1):

$$M_{c_j-\varepsilon} \cup D^i \times D^{m-i}.$$

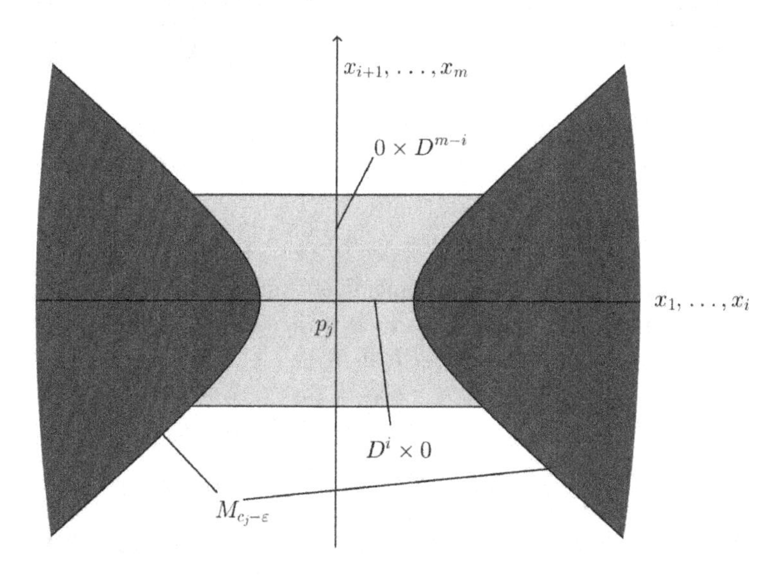

Fig. 3.1 An i-handle

Note that this space is not a manifold, since there are "corners" where the i-handle meets the sublevel set $M_{c_j-\varepsilon}$. However, we can smooth this space out to form a manifold M' which retracts onto $M_{c_j-\varepsilon} \cup D^i \times D^{m-i}$. We will show that this space M' is a retract of $M_{c_j+\varepsilon}$, and since M' retracts onto $M \cup (D^i \times 0)$, we see that $M_{c_j+\varepsilon}$ has the homotopy type of $M_{c_j-\varepsilon} \cup (D^i \times 0)$.

Theorem 3.3. *The sublevel set $M_{c_j+\varepsilon}$ is diffeomorphic to the smoothed manifold M'.*

Proof. A quick heuristic is provided by the proof of Theorem 3.1: use a gradient-like vector field for f to let $\partial M'$ flow upward to $\partial M_{c_j+\varepsilon}$. This gives us the required diffeomorphism, and flowing backwards gives the deformation retraction. See Figure 3.2 for an illustration.

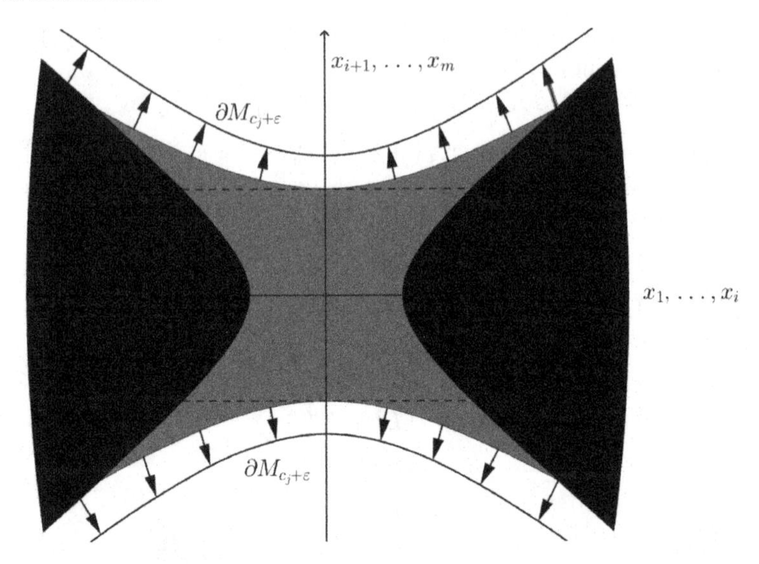

Fig. 3.2 The manifold M' (the entire shaded region) obtained from attaching an i-handle to $M_{c_j-\varepsilon}$ (the black region) and then smoothing the corners. The region inside the hyperbola crossing the $x_{i+1}, \ldots, x_m$-axis is the entire sublevel set $M_{c_j+\varepsilon}$ and the arrows show the flow outward from M' along a gradient-like vector field for f.

The actual proof is much more technical (cf. [Milnor (1965b)]). We may assume our coordinate neighborhood U around p_j is such that the image under the embedding $U \to \mathbb{R}^m$ contains the closed ball

$$\{(x_1, \ldots, x_m) : \sum_{i=1}^{m} x_i^2 \leq 2\varepsilon\}.$$

Note that the core $D^i \times 0$ is then contained in U. Note also that $(D^i \times 0) \cap M_{c_j-\varepsilon}$ is the boundary $S^{i-1} \times 0$ so that the core is attached to $M_{c_j-\varepsilon}$ along that boundary sphere. We claim that $M_{c_j-\varepsilon} \cup (D^i \times 0)$ is a deformation retract of $M_{c_j+\varepsilon}$. Along the way, we will construct M', which lies between the two, and see that it is diffeomorphic to $M_{c_j+\varepsilon}$.

Choose a smooth function $\mu : \mathbb{R} \to \mathbb{R}$ satisfying the following conditions:

$$\mu(0) > \varepsilon$$
$$\mu(t) = 0 \text{ for } t \geq 2\varepsilon$$
$$\mu'(t) \in (-1, 0] \text{ for all } t.$$

Define a new function $F : M \to \mathbb{R}$ by taking F to equal the original function f outside the open set U and setting

$$F = f - \mu(x_1^2 + \cdots + x_i^2 + 2x_{i+1}^2 + \cdots 2x_m^2)$$

inside U. It is easy to see that F is a smooth function on all of M. For notational convenience, define $\xi, \eta : U \to [0, \infty)$ by

$$\xi = x_1^2 + \cdots + x_i^2$$
$$\eta = x_{i+1}^2 + \cdots + x_m^2.$$

Then $f = c_j - \xi + \eta$ and hence for any $z \in U$

$$F(z) = c_j - \xi(z) + \eta(z) - \mu(\xi(z) + 2\eta(z)).$$

The first thing to note is that $F^{-1}(-\infty, c_j + \varepsilon] = M_{c_j+\varepsilon}$. Indeed, outside of the region $\xi + 2\eta \leq 2\varepsilon$ the functions f and F agree, while within this region

$$F \leq f = c_j - \xi + \eta \leq c_j + \frac{1}{2}\xi + \eta \leq c_j + \varepsilon.$$

Furthermore, we claim that the critical points of F agree with those of f. To see this note that

$$\frac{\partial F}{\partial \xi} = -1 - \mu'(\xi + 2\eta) < 0$$
$$\frac{\partial F}{\partial \eta} = 1 - 2\mu'(\xi + 2\eta) \geq 1$$

and since

$$dF = \frac{\partial F}{\partial \xi} d\xi + \frac{\partial F}{\partial \eta} d\eta$$

vanishes only at the origin in U (as $d\xi$ and $d\eta$ vanish simultaneously only at that point), F has no critical points in U except for the origin.

Now, since $F^{-1}(-\infty, c_j + \varepsilon] = M_{c_j+\varepsilon}$ and since $F \leq f$, we have an inclusion

$$F^{-1}[c_j - \varepsilon, c_j + \varepsilon] \subset f^{-1}[c_j - \varepsilon, c_j + \varepsilon],$$

and thus $F^{-1}[c_j - \varepsilon, c_j + \varepsilon]$ is compact. Denote the sublevel set $F^{-1}(-\infty, c_j - \varepsilon]$ by M'. By Remark 3.2, we then deduce that M' is a deformation retract of $M_{c_j+\varepsilon}$, and that the two spaces are diffeomorphic.

Write $M' = M_{c_j-\varepsilon} \cup H$, where H is the closure of $M' - M_{c_j-\varepsilon}$. We will now show that $M_{c_j-\varepsilon} \cup (D^i \times 0)$ is a deformation retract of M'. Note, however, that this retraction is not a diffeomorphism as the set H has dimension m, while the core has dimension i. Recall that the core $D^i \times 0$ consists of points z with $\xi(z) \leq \varepsilon$ and $\eta(z) = 0$. It follows that $D^i \times 0 \subset H$. The retraction can be seen by examining Figure 3.2. The handle is the lightly shaded region and the idea is to push into the core along the gradient of f. Explicitly, we define a retraction $r : M' \times [0, 1] \to M'$ in steps. Set $r(z, t)$ to be the identity outside of the coordinate neighborhood U. First suppose $(x_1, \ldots, x_m)$ lies in the region $\xi \leq \varepsilon$. This corresponds to the region between the vertical lines passing through the vertices of the black hyperbola in Figure 3.2. Set

$$r((x_1, \ldots, x_m), t) = (x_1, \ldots, x_i, tx_{i+1}, \ldots, tx_m).$$

Then $r(z,1) = z$ and $r(z,0)$ lies in the core $D^i \times 0$ for all z. That $r(-,t)$ maps M' into itself follows from the fact that $\dfrac{\partial F}{\partial \eta} > 0$.

If z lies in the region $\varepsilon \leq \xi \leq \eta + \varepsilon$ (these are the points in H that lie between the black hyperbola and the gray hyperbola, but outside the vertical lines passing through the black hyperbola's vertices), then set

$$r((x_1, \ldots, x_m), t) = (x_1, \ldots, x_i, s_t x_{i+1}, \ldots, s_t x_m)$$

where

$$s_t = t + (1-t)\sqrt{\frac{t-\varepsilon}{\eta}}.$$

Note that $r(-,1)$ is the identity and $r(-,0)$ maps the entire region into $M_{c_j - \varepsilon}$. We leave it as an exercise to check that the functions $s_t x_\ell$ are continuous as $\xi \to \varepsilon$ and $\eta \to 0$. Note also that this definition agrees with that in the first case when $\xi = \varepsilon$.

Finally, if z lies in $M_{c_j - \varepsilon}$ (i.e., within the region $\eta + \varepsilon \leq \xi$) we can take $r(-,t)$ to be the identity. This agrees with the previous case when $\xi = \eta + \varepsilon$. This shows that M' retracts onto $M_{c_j - \varepsilon} \cup H$ and completes the proof of the theorem. □

We now understand what happens to the topology of the sublevel sets as we pass critical points of every possible index: passing a critical point of index i results in attaching an i-handle to the existing sublevel set along its boundary. That is, we must specify an *attaching map*

$$\psi : \partial D^i \times D^{m-i} \to \partial M_{c_j - \varepsilon}$$

and then we glue the i-handle to $M_{c_j - \varepsilon}$ using this function. The boundary ∂D^i of the core disc is an $(i-1)$-sphere S^{i-1} called the *attaching sphere*. This leads us to the following definition.

Definition 3.4. An m-dimensional handlebody is a manifold (possibly with boundary) defined as follows.

(1) A disc D^m is an m-dimensional handlebody.
(2) If $\psi_1 : \partial D^{i_j} \times D^{m-i_j} \to \partial D^m$ is an attaching map, then the manifold

$$D^m \cup_{\psi_1} D^{i_j} \times D^{m-i_j}$$

is an m-dimensional handlebody, denoted by $H(D^m; \psi_1)$.
(3) If $M' = H(D^m; \psi_1, \ldots, \psi_{k-1})$ is an m-dimensional handlebody, and if $\psi_k : \partial D^{i_\ell} \times D^{m-i_\ell} \to \partial M'$ is an attaching map, then

$$M' \cup_{\psi_k} D^{i_\ell} \times D^{m-i_\ell}$$

is an m-dimensional handlebody, denoted $H(D^m; \psi_1, \ldots, \psi_k)$.

With this definition in hand we may now prove the following result.

Theorem 3.5. *Suppose $f : M \to \mathbb{R}$ is a Morse function on the smooth, compact m-manifold M. Then f determines a handlebody structure on M, with an i-handle for each index-i critical point of f.*

Proof. We may assume that f has distinct critical values; order the critical points by ascending f-values: $p_0, p_1, \ldots, p_k$. We proceed by induction, the base case following from the fact that since p_0 is the absolute minimum of f the sublevel set $M_{c_0+\varepsilon}$ is an m-disc D^m and is therefore an m-dimensional handlebody.

Now assume that $M_{c_{j-1}+\varepsilon}$ is a handlebody $H(D^m; \psi_1, \ldots, \psi_{j-1})$. We know that $M_{c_j+\varepsilon}$ is obtained from $M_{c_j-\varepsilon}$ by attaching an i_j-handle. The attaching map $\psi : \partial D^{i_j} \times D^{m-i_j} \to M_{c_j-\varepsilon}$ is determined by f. Since $[c_{j-1}+\varepsilon, c_j - \varepsilon]$ contains no critical values of f, Theorem 3.1 asserts that $M_{c_{j-1}+\varepsilon}$ is diffeomorphic to $M_{c_j-\varepsilon}$. Let Ψ be a diffeomorphism obtained by choosing a gradient-like vector field for f and letting $M_{c_{j-1}+\varepsilon}$ flow along the integral lines.

By the inductive hypothesis, we know that $M_{c_j-\varepsilon}$ is diffeomorphic to the handlebody $H(D^m; \psi_1, \ldots, \psi_{j-1})$. But we know by Theorem 3.3 that $M_{c_j+\varepsilon}$ is obtained from this space by attaching an i_j-handle and is therefore an m-dimensional handlebody. This completes the inductive step and the theorem is proved. □

We should take some care to be more explicit about the attaching map of the i_j-handle in the inductive step. The map Ψ is a diffeomorphism $H(D^m; \psi_1, \ldots, \psi_{j-1}) \to M_{c_j-\varepsilon}$. Now, the attaching map for the handle maps to the latter space, not the handlebody, and the attaching map, call it ψ, is determined by f. The attaching map for the handlebody is then $\Psi^{-1} \circ \psi$ and denoting this composite by ψ_j, we then obtain a handlebody

$$H(D^m; \psi_1, \ldots, \psi_{j-1}) \cup_{\psi_j} D^{i_j} \times D^{m-i_j} = H(D^m; \psi_1, \ldots, \psi_j).$$

Note also that a handlebody decomposition depends on the function f. That is, a manifold M admits many different handlebody decompositions, one for each Morse function on M. Even for a fixed function f, the structure of the handle decomposition can vary with the choice of a gradient-like vector field for f.

From a topological point of view, it is often more convenient to think in terms of homotopy type. We showed in Theorem 3.3 that the manifold M is built by attaching cores. That is, any given sublevel set $M_{c_j+\varepsilon}$ retracts on to $M_{c_j-\varepsilon} \cup (D^i \times 0)$, where i is the index of the critical point p_j. It follows that we may build a CW complex of the same homotopy type as M by beginning with a 0-cell corresponding to the global minimum of f, and attaching cells of dimension i for each critical point of index i (see Appendix B for more information about CW complexes).

Theorem 3.6. *Let M be a smooth compact m-manifold and let $f : M \to \mathbb{R}$ be a Morse function with critical points $p_0, p_1, \ldots, p_k$. Denote the index of p_j by i_j. Then M has the homotopy type of a CW-complex with one cell of dimension i_j for each critical point p_j; that is,*

$$M \simeq e^{i_0} \cup e^{i_1} \cup \cdots \cup e^{i_k}$$

where e^ℓ denotes a cell of dimension ℓ.

Proof. We will need the following two facts, proofs of which may be found in [Milnor (1963)], Section 3.

(1) Suppose $f, g : \partial e^n \to X$ are homotopic maps. Then the identity map of X extends to a homotopy equivalence $F : X \cup_f e^n \to X \cup_g e^n$.
(2) Suppose $f : \partial e^n \to X$ is an attaching map. Then any homotopy equivalence $g : X \to Y$ extends to a homotopy equivalence $G : X \cup_f e^n \to Y \cup_{g \circ f} e^n$.

Given these, let us proceed. Let $c_0 < c_1 < \cdots < c_k$ be the critical values of f. For $a < c_0$, we have $M_a = \emptyset$ and if ε is small, we have $M_{c_0+\varepsilon} = D^m$, which has the homotopy type of a 0-cell e^0. Now, suppose that $a \neq c_i$ and that M_a has the homotopy type of a CW-complex. Say c_i is the smallest critical value greater than a. We know that $M_{c+\varepsilon}$ has the homotopy type of $M_{c-\varepsilon} \cup_\psi e^i$ and that there is a homotopy equivalence $h : M_{c-\varepsilon} \to M_a$. Our inductive hypothesis tells us that there is a homotopy equivalence $k : M_a \to K$, where K is a CW-complex. By cellular approximation, the map $k \circ h \circ \psi$ is homotopic to a map

$$\varphi : \partial e^i \to K^{(i-1)},$$

where $K^{(i-1)}$ denotes the $(i-1)$-skeleton of K. Thus $K \cup_\varphi e^i$ is a CW-complex and has the same homotopy type as $M_{c+\varepsilon}$ by the two facts stated above. This completes the inductive step and the proof is complete. □

3.2 Examples of handlebody decompositions

Example 3.7. The m-sphere. Let S^m denote the unit sphere in $\mathbb{R}^{m+1}$:

$$S^m = \{(x_1, \ldots, x_{m+1}) : x_1^2 + \cdots + x_{m+1}^2 = 1\}.$$

The height function

$$f(x_1, \ldots, x_{m+1}) = x_{m+1}$$

is a Morse function with exactly two critical points: the south pole, of index 0, and the north pole, of index m. It follows that S^m has a handle decomposition with one 0-handle and one m-handle:

$$S^m = D^m \cup D^m,$$

which corresponds to the decomposition of the sphere into northern and southern hemispheres. The attaching map $\psi : \partial D^m \to \partial D^m$ is the identity map on the boundary spheres. That is, we may take $\varepsilon = 1$ so that the sublevel set $S^m_{-1+\varepsilon}$ is the 0-handle D^m and we glue the northern hemisphere to it along the equator (which is the level set $f^{-1}(0)$).

Example 3.8. Real projective m-space. Denote by $\mathbb{R}P^m$ the set of lines through the origin in the Euclidean space $\mathbb{R}^{m+1}$. A nonzero point $(x_1, \ldots, x_{m+1})$ in $\mathbb{R}^{m+1}$ determines a unique line through the origin and hence gives a point in $\mathbb{R}P^m$, denoted $[x_1 : x_2 : \cdots : x_{m+1}]$. Note that two such lines, one passing through $(x_1, \ldots, x_{m+1})$

and another through $(y_1, \ldots, y_{m+1})$, are the same precisely when there is an $\alpha \neq 0$ with

$$(y_1, \ldots, y_{m+1}) = (\alpha x_1, \ldots, \alpha x_{m+1}).$$

The two points in $\mathbb{R}P^m$ then agree:

$$[y_1 : \cdots : y_{m+1}] = [x_1 : \cdots : x_{m+1}].$$

At first glance it is not at all clear how to put a topology on this set, nor how to give it the structure of a manifold. There is a natural topology, but it is simpler to express $\mathbb{R}P^m$ as a quotient space and give it the induced quotient topology. This has the added advantage of allowing us to deduce that $\mathbb{R}P^m$ is compact.

To this end, define a map $\pi : S^m \to \mathbb{R}P^m$ as follows. Any point $(x_1, \ldots, x_{m+1}) \in S^m$ determines a unique line through the origin. Set $\pi(x_1, \ldots, x_{m+1}) = [x_1 : \cdots : x_{m+1}]$. This map is surjective, and giving $\mathbb{R}P^m$ the quotient topology induced by π, we get continuity for free. Since S^m is compact and $\mathbb{R}P^m$ is its continuous image under π, we see that $\mathbb{R}P^m$ is compact as well.

Observe that π is a $2:1$ map; that is, it maps two points in S^m to any given point $[y_1 : \cdots : y_{m+1}]$. Indeed, we may assume that the components of $[y_1 : \cdots : y_{m+1}]$ satisfy the condition $y_1^2 + \cdots + y_{m+1}^2 = 1$ and then π takes $(y_1, \ldots, y_{m+1})$ and its antipode $(-y_1, \ldots, -y_{m+1})$ to $[y_1 : \cdots : y_{m+1}]$.

As a simple example, consider $\mathbb{R}P^1$, which consists of the lines through the origin in $\mathbb{R}^2$. We claim that this space is simply S^1. Indeed, any line through the origin in $\mathbb{R}^2$ cuts the unit circle in exactly two points. It follows that, as a set, $\mathbb{R}P^1$ consists of the upper semicircle but with the points $(1,0)$ and $(-1,0)$ identified. This space is clearly just S^1. The spaces $\mathbb{R}P^m$ for $m > 1$ are much more complicated, of course, and if m is even they are nonorientable manifolds.

Let us now deduce a handlebody structure on $\mathbb{R}P^m$. To this end, we need to show that $\mathbb{R}P^m$ admits the structure of an m-manifold and then find a suitable Morse function on it. Fix i and consider the set U_i of points in $\mathbb{R}P^m$ consisting of those $[x_1 : \cdots : x_{m+1}]$ for which $x_i \neq 0$. This is an open set in $\mathbb{R}P^m$ as its inverse image under π consists of the two open hemispheres complementary to the "equator" $x_i = 0$ in S^m. Moreover, since each of these open hemispheres is an open m-ball, we see that U_i is diffeomorphic to an m-ball (restrict π to one hemisphere) and thus we may take the U_i as a system of coordinate neighborhoods on $\mathbb{R}P^m$. A convenient coordinate system $(X_1, \ldots, X_m)$ on U_i is given by

$$X_1 = \frac{x_1}{x_i}, \ldots, X_{i-1} = \frac{x_{i-1}}{x_i}, X_i = \frac{x_{i+1}}{x_i}, \ldots, X_m = \frac{x_{m+1}}{x_i}.$$

Choose real numbers $a_1 < a_2 < \cdots < a_{m+1}$ and define $f : \mathbb{R}P^m \to \mathbb{R}$ by

$$f([x_1 : \cdots : x_{m+1}]) = \frac{a_1 x_1^2 + \cdots + a_{m+1} x_{m+1}^2}{x_1^2 + \cdots + x_{m+1}^2}.$$

Note that this is well-defined since multiplying all the x_j by a fixed $\alpha \neq 0$ does not affect the value on the right-hand side above. Now, fix an i and consider f in the coordinate system $(X_1, \ldots, X_m)$ on U_i:

$$f(X_1, \ldots, X_m) = \frac{a_1 X_1^2 + \cdots + a_{i-1} X_{i-1}^2 + a_i + a_{i+1} X_i^2 + \cdots + a_{m+1} X_m^2}{X_1^2 + \cdots + X_{i-1}^2 + 1 + X_i^2 + \cdots + X_m^2}.$$

It is easy to check, by first differentiating with respect to X_m, then X_{m-1}, ..., then X_i, and using the fact that $a_{m+1} > a_m > \cdots > a_{i+1}$ that any critical point of f must satisfy

$$X_i = \cdots = X_{m-1} = X_m = 0.$$

Then, differentiating with respect to X_1, then X_2, ... then X_{i-1} and noting that $a_1 < a_2 < \cdots < a_i$, we see that any critical point has

$$X_1 = X_2 = \cdots = X_{i-1} = 0.$$

Thus, the only critical point of f in U_i is the origin $(0, \ldots, 0)$, which corresponds to the point $p_i = [0 : \cdots : 0 : 1 : 0 : \cdots : 0]$ in $\mathbb{R}P^m$ (the 1 is the ith entry).

A quick calculation shows that the Hessian at p_i is the diagonal matrix with entries $2(a_1 - a_i), \ldots, 2(a_{i-1} - a_i), 2(a_{i+1} - a_i), \ldots, 2(a_{m+1} - a_i)$. It follows that the index of p_i is $i - 1$. Note that $f(p_i) = a_i$. Thus, f has $m + 1$ critical points of indices

$$0, 1, \ldots, m$$

in ascending order. It follows that we have a handlebody decomposition

$$\mathbb{R}P^m = D^m \cup (D^1 \times D^{m-1}) \cup (D^2 \times D^{m-2}) \cup \cdots \cup (D^{m-1} \times D^1) \cup D^m.$$

Note that when $m = 1$, we get $\mathbb{R}P^1 = D^1 \cup D^1 \cong S^1$.

We also get a CW-complex of the homotopy type of $\mathbb{R}P^m$ thanks to Theorem 3.6. We begin with a 0-cell, then attach a 1-cell via the only possible map. The 2-cell is attached via a map $\psi_2 : \partial D^2 \to \mathbb{R}P^1$; it is easily seen that this is the double cover $\pi : S^1 \to \mathbb{R}P^1$. In general, we obtain $\mathbb{R}P^m$ from $\mathbb{R}P^{m-1}$ by attaching an m-cell via the double cover $\pi : S^{m-1} \to \mathbb{R}P^{m-1}$. The resulting cell decomposition is

$$\mathbb{R}P^m = e^0 \cup e^1 \cup \cdots \cup e^m.$$

Example 3.9. The special orthogonal group $SO(m)$. Recall that an $m \times m$ matrix A is *orthogonal* if each column is a unit vector in $\mathbb{R}^m$ and if the columns are mutually orthogonal. An orthogonal matrix A with $\det A = 1$ is called a rotation matrix and the set of all such matrices is denoted by $SO(m)$. This set forms a group under matrix multiplication, called the *special orthogonal group*. Moreover, $SO(m)$ is a smooth manifold and since it is a closed subset of the product $S^{m-1} \times \cdots \times S^{m-1}$, it is compact.

Observe that $SO(1)$ is trivial. The group $SO(2)$ is diffeomorphic to the unit circle S^1. Indeed, any matrix in $SO(2)$ has the form

$$A(\theta) = \begin{pmatrix} \cos\theta & -\sin\theta \\ \sin\theta & \cos\theta \end{pmatrix}$$

and the map $A(\theta) \mapsto \theta$ gives the required diffeomorphism. In general, the manifold $SO(m)$ has dimension

$$(m-1)+(m-2)+\cdots+2+1=\frac{m(m-1)}{2}.$$

This follows by noting that we have an entire S^{m-1} of choices for the first column, and then an orthogonal S^{m-2} to the first choice, and so on. Now, fix real numbers $1 < c_1 < c_2 < \cdots < c_m$ and define a map $f : SO(m) \to \mathbb{R}$ by

$$f(X) = c_1 x_{11} + c_2 x_{22} + \cdots + c_m x_{mm}.$$

This map is clearly smooth.

Lemma 3.10. *The critical points of f consist precisely of those diagonal matrices with entries ± 1, chosen arbitrarily as long as the determinant is* 1.

Proof. Note that we may embed $SO(2)$ into $SO(m)$ as the upper left 2×2 minor. Let $A(\theta)$ denote the matrix defined above, viewed as an element of $SO(m)$. The product of an arbitrary X with $A(\theta)$ gives a curve in $SO(m)$ with parameter θ, passing through X at $\theta = 0$. Computing this product and then evaluating at f, we see that

$$f(XA(\theta)) = c_1(x_{11}\cos\theta + x_{12}\sin\theta) + c_2(-x_{21}\sin\theta + x_{22}\cos\theta) + c_3 x_{33} + \cdots c_m x_{mm}.$$

Differentiating with respect to θ and setting $\theta = 0$, we find

$$\left.\frac{d}{d\theta} f(XA(\theta))\right|_{\theta=0} = c_1 x_{12} - c_2 x_{21}.$$

Repeating this for the curve $A(\theta)X$, we find

$$\left.\frac{d}{d\theta} f(A(\theta)X)\right|_{\theta=0} = -c_1 x_{21} + c_2 x_{12}.$$

Now, if X is a critical point of f, then both of these derivatives must be 0. Since $1 < c_1 < c_2$, it follows that $x_{12} = x_{21} = 0$.

Now, by embedding $SO(2)$ via the k, ℓ-coordinate planes (rather than the 1, 2-plane), we may repeat this calculation to find that $x_{k\ell} = x_{\ell k} = 0$ for all $k \neq \ell$. Thus, X is a diagonal matrix and must have the form in the statement of the lemma.

It remains to show that such matrices are actually critical points of f. Let $X = \operatorname{diag}(x_1, \ldots, x_m)$ be such a matrix; that is, each $x_j = \pm 1$ and $\det X = 1$. Repeating the above calculation for the curve $XA(\theta)$ shows that the tangent vector at X along this curve is a matrix consisting of all zeroes except that the 1, 2-entry is $-x_1$ and the 2, 1-entry is x_2. Similarly, embedding $SO(2)$ in $SO(m)$ in the k, ℓ-plane (call the resulting matrices $A_{k\ell}(\theta)$ so that the $A(\theta)$ introduced above is $A_{12}(\theta)$) yields a matrix with k, ℓ-entry $-x_i$ and ℓ, k-entry x_k ($k < \ell$, here). Call this matrix $U_{k\ell}$. These are linearly independent vectors in the vector space of all $m \times m$ matrices and there are $\dfrac{m(m-1)}{2}$ of them. Since this is the dimension of $SO(m)$, they must form a basis of the tangent space $T_X SO(m)$.

Finally, if X is one of the diagonal matrices with entries ± 1, it is an easy calculation to show that the derivative of f in the direction of each $U_{k\ell}$ vanishes. It follows that X is a critical point of f. $\square$

Table 3.1 The critical points for $SO(4)$ and their indices

critical point	index
$(-1,-1,-1,-1)$	0
$(-1,-1,1,1)$	1
$(-1,1,-1,1)$	2
$(-1,1,1,-1)$	3
$(1,-1,-1,1)$	3
$(1,-1,1,-1)$	4
$(1,1,-1,-1)$	5
$(1,1,1,1)$	6

We leave it as an exercise to show that each of these critical points is nondegenerate (indeed, the matrix of second partial derivatives with respect to an appropriate coordinate system is diagonal with nonzero entries). If X is one of these matrices, with diagonal entries $x_1, \ldots, x_m$, the k, ℓ entry of the Hessian matrix is $-c_k x_k - c_\ell x_\ell$. Assume that the entries with $x_i = 1$ have subscripts $i_1, \ldots, i_r$. Then the number of minus signs on the diagonal is

$$(i_1 - 1) + (i_2 - 1) + \cdots + (i_r - 1).$$

The index is 0 if all the $x_i = -1$. The critical values are all distinct provided $2c_i < c_{i+1}$ for all $i = 1, \ldots, m-1$. Observe that there are 2^{m-1} critical points because of the condition $\det X = 1$.

For example, when $m = 2$, we get two critical points: $X_1 = \text{diag}(-1,-1)$ of index 0 and $X_2 = \text{diag}(1,1)$ of index $(1-1)+(2-1) = 1$. This recovers the diffeomorphism $SO(2) \cong S^1$. The example of $SO(3)$ will be treated in the exercises. Note that $SO(4)$ has 8 critical points with indices as in Table 3.1.

Example 3.11. The unitary group $U(m)$. If $Z = (z_{ij})$ is a complex $m \times m$ matrix, denote by Z^* its conjugate transpose: $Z^* = (\overline{z_{ji}})$. The set $U(m)$ of all matrices satisfying $Z^* Z = I$ forms a complex Lie group under matrix multiplication called the *unitary group*. Note that a matrix $Z \in U(m)$ satisfies $|\det Z| = 1$. An alternate definition of $U(m)$ is as the set of all matrices $Z = (\mathbf{z}_1, \ldots, \mathbf{z}_m)$ whose columns satisfy $\mathbf{z}_i \cdot \overline{\mathbf{z}_i} = 1$ and $\mathbf{z}_i \cdot \overline{\mathbf{z}_j} = 0$, $i \neq j$.

To compute the dimension of $U(m)$, note that we may choose any vector of unit length in $\mathbb{C}^m$ for the first column $\mathbf{z}_1$. The set of such is the unit sphere in $\mathbb{C}^m$; this has real dimension $(2m-1)$. The second vector may then be any unit vector in the orthogonal $(m-1)$-dimensional complex subspace, contributing another $2(m-1)-1$ dimensions. It follows that

$$\dim_{\mathbb{R}} U(m) = (2m-1) + (2m-3) + \cdots + 3 + 1 = m^2.$$

Now, given real numbers $1 < c_1 < \cdots < c_m$ define a function $f : U(m) \to \mathbb{R}$ by

$$f(Z) = \text{Re}(c_1 z_{11} + c_2 z_{22} + \cdots + c_m z_{mm}).$$

This is a smooth function and using arguments similar to those for $SO(m)$ we can deduce a handlebody structure for $U(m)$. We need a basis for the tangent space at

Table 3.2 The critical points for $U(2)$ and their indices

critical point	index
$(-1,-1)$	0
$(1,-1)$	1
$(-1,1)$	3
$(1,1)$	4

any $Z \in U(m)$. To that end, we use the real rotation matrices $A_{k\ell}(\theta)$ in $SO(m)$, their complex analogues $B_{k\ell}(\theta)$ where $B(\theta) \in U(2)$ is

$$B(\theta) = \begin{pmatrix} \cos\theta & i\sin\theta \\ i\sin\theta & \cos\theta \end{pmatrix},$$

and the diagonal matrices $C_{kk}(\theta) = \mathrm{diag}(1, 1, \ldots, e^{i\theta}, \ldots, 1)$, where $e^{i\theta}$ lies in the kth position. Using the same argument as in the case of $SO(m)$, we see that these m^2 rotation directions provide a basis for the tangent space of $U(m)$. Moreover, the same calculation shows that the critical points of $f : U(m) \to \mathbb{R}$ are the matrices with diagonal entries ± 1, without the restriction that the matrix have determinant 1. We therefore find 2^m critical points $\mathrm{diag}(\pm 1, \pm 1, \ldots, \pm 1)$. If $i_1, \ldots, i_r$ are those subscripts of a critical matrix having entry 1, then the index of the critical point is

$$(2i_1 - 1) + (2i_2 - 1) + \cdots + (2i_r - 1).$$

The index is 0 if all the $x_i = -1$.

As an example, consider $U(2)$. The four critical points and their indices are shown in Table 3.2. Note that the handle dimensions agree with those of the product manifold $S^1 \times S^3$. There is a general result in this direction; it is explored in the exercises.

3.3 Sliding and canceling handles

Knowledge of the indices of the critical points of a Morse function is not sufficient to determine the diffeomorphism type of a manifold; we must understand the attaching maps. We have met $\mathbb{R}P^2$, which is obtained from an annulus (a 0-handle with a 1-handle attached) by attaching a 2-handle via the double cover $\pi : S^1 \to S^1$, but there are infinitely many homotopy classes of maps $S^1 \to S^1$, each yielding a distinct handlebody when used as an attaching map. We therefore must understand how we may alter an attaching map without changing the manifold.

Suppose $f : M \to \mathbb{R}$ is a Morse function on the compact m-manifold M and let X be a gradient-like vector field for f. Assume the critical values of f are distinct and that the critical points $p_0, p_1, \ldots, p_k$ are arranged by increasing order of the critical values. By Theorem 3.5 we have a handlebody decomposition

$$M = D^m \cup_{\psi_1} (D^{i_1} \times D^{m-i_1}) \cup_{\psi_2} \cdots \cup_{\psi_{k-1}} (D^{i_{k-1}} \times D^{m-i_{k-1}}) \cup_{\psi_k} D^m.$$

Denote by M_j the subhandlebody $H(D^m; \psi_1, \ldots, \psi_j)$ obtained by attaching the 0-th handle through the jth. Then the ℓth handle is attached to $M_{\ell-1}$ via $\psi_\ell : \partial D^{i_\ell} \times D^{m-i_\ell} \to \partial M_{\ell-1}$.

If N is a smooth manifold, recall that an *isotopy* is a collection $\{f_t\}_{t\in J}$ (J an open interval containing $[0,1]$) such that each $f_t : N \to N$ is a diffeomorphism, $f_t = \mathrm{id}_N$ for $t \leq 0$, $f_t = f_1$ for $t \geq 1$, and the map $F : N \times J \to N \times J$ defined by $F(n,t) = (f_t(n), t)$ is a diffeomorphism. We now consider the following question: given an isotopy of ∂M_j, how does the diffeomorphism type of the subhandlebodies M_ℓ change?

Theorem 3.12. *Suppose $\{f_t\}_{t\in J}$ is an isotopy of the boundary ∂M_{j-1} of the subhandlebody M_{j-1}. Let $\varphi_j = f_1 \circ \psi_j$. Then the attaching map $\psi_j : \partial D^{i_j} \times D^{m-i_j} \to \partial M_{j-1}$ can be replaced by φ_j without altering the diffeomorphism type of each M_ℓ, $0 \leq \ell \leq k$.*

Proof. The Morse function f and gradient-like vector field X determine the handlebody decomposition

$$M = D^m \cup_{\psi_1} D^{i_1} \times D^{m-i_1} \cup_{\psi_2} \cdots \cup_{\psi_{k-1}} D^{i_{k-1}} \times D^{m-i_{k-1}} \cup_{\psi_k} D^m.$$

Fix j. The sublevel set $M_{c_j+\varepsilon}$ is diffeomorphic to the subhandlebody $M_j = H(D^m; \psi_1, \ldots, \psi_j)$. Consider the ℓth handle. The attaching map is determined by a map $\psi : \partial D^{i_\ell} \times D^{m-i_\ell} \to \partial M_{c_\ell - \varepsilon}$. We also have a diffeomorphism $\Psi : M_{c_{\ell-1}+\varepsilon} \to M_{c_\ell-\varepsilon}$ determined by the gradient-like vector field X. Identifying $N_{\ell-1}$ with the domain of Ψ, the attaching map ψ_i is the map $\Psi^{-1} \circ \psi$.

Before proceeding, we note that $f^{-1}([c_{\ell-1}+\varepsilon, c_\ell - \varepsilon]) \cong \partial M_{c_{\ell-1}+\varepsilon} \times [0,1]$. This map extends to a slightly larger interval and there exists a sufficiently small $\delta > 0$ so that

$$f^{-1}([c_{\ell-1}+\varepsilon/2, c_\ell - \varepsilon/2]) \cong \partial M_{c_{\ell-1}+\varepsilon} \times [-\delta, 1+\delta],$$

the diffeomorphism still given by the integral curves of X. The map Ψ is then really just the diffeomorphism that streches $\partial M_{c_{\ell-1}+\varepsilon} \times [-\delta, 0]$ to $\partial M_{c_{\ell-1}+\varepsilon} \times [-\delta, 1]$.

The identification of $M_{\ell-1}$ with $M_{c_{\ell-1}+\varepsilon}$ allows us to view the isotopy $\{f_t\}$ on $\partial M_{c_{\ell-1}+\varepsilon}$. The map $F : \partial M_{c_{\ell-1}+\varepsilon} \times J \to \partial M_{c_{\ell-1}+\varepsilon} \times J$ defined by $F(x,t) = (f_t(x), t)$ is a diffeomorphism, independent of t for $t \leq 0$ or $t \geq 1$. Set $G(x,t) = (f_{1-t}(x), t)$, the map F in reverse time. The map G is also a diffeomorphism. Identify J with the interval $(-\delta, 1+\delta)$. Now, the integral curves of X in $\partial M_{c_{\ell-1}+\varepsilon} \times (-\delta, 1+\delta)$ are $\{x\} \times (-\delta, 1+\delta)$ and so X can be identified with the vector field $\dfrac{\partial}{\partial t}$. The map G induces a vector field $G_*(X)$, and since G is constant for $t \leq 0$ and $t \geq 1$, $G_*(X)$ equals X on these regions. So, replacing X by $G_*(X)$ in $\partial M_{c_{\ell-1}+\varepsilon} \times (-\delta, 1+\delta)$, the new vector field $G_*(X)$ extends smoothly outside this region to the original X. Call this new vector field Y; it is a gradient-like vector field for f.

The vector field Y determines a diffeomorphism

$$\Xi : M_{c_{\ell-1}+\varepsilon} \to M_{c_\ell-\varepsilon}$$

analogous to the map Ψ. In the handle decomposition determined by f and Y, the ℓth handle is attached to $M_{c_{\ell-1}+\varepsilon}$ via the map

$$\Xi \circ \psi = f_1 \circ \psi_\ell.$$

The handlebodies up to the $(\ell-1)$th are unchanged since X and Y coincide there. Moreover since the definition of $M_{c_j+\varepsilon}$ does not depend on the gradient-like vector field, the diffeomorphism type of M_j remains unchanged for all j.

□

The proof of Theorem 3.12 is very technical, but the idea is very simple. What this theorem really says is that the gradient-like vector field X can be perturbed to another gradient-like vector field Y so that the associated handlebody decomposition has the properties that (i) the structure of the handlebody decompositions up through the $(j-1)$st are unchanged; and (ii) if M'_ℓ denotes the new subhandlebody, then for any ℓ the diffeomorphism type of M'_ℓ is the same as that as M_ℓ. The picture to keep in mind is that we are attaching the boundary of a handle in a new position on $\partial M_{\ell-1}$; that is, we are "sliding" the handle on the boundary. The theorem asserts that this does not change the diffeomorphism type of M, although it does change the handlebody decomposition.

Theorem 3.6 tells us that a manifold M has the homotopy type of a CW-complex with a cell of dimension i for each critical point of index i. When one builds such a complex inductively, it is common to attach cells in increasing order of dimension; that is, one attaches an i-cell only after all the cells of dimension $< i$ have already been added. This is not strictly necessary, of course. Handlebody decompositions do not necessarily proceed in this manner. Indeed, one often passes a local minimum of the Morse function, resulting in the addition of a disjoint m-disc. We would like to be able to attach handles in increasing index order. To this end we will need some new definitions and a few lemmas.

Lemma 3.13. *Let A and B be compact submanifolds of M of dimensions a and b, respectively. If $a+b<m$, where $m=\dim M$, then there is an isotopy $\{f_t\}_{t\in J}$ of M such that $f_0 = id_M$ and $f_1(A)\cap B=\emptyset$.*

Proof. This is a standard result in manifold theory. The example to keep in mind is that of two lines in $\mathbb{R}^3$. Two such lines probably do not intersect anyway, but if they do, it is clearly possible to move one to make it disjoint from the other. We omit the proof; one may be found in e.g., [Matsumoto (1997)], p. 111. □

Now, suppose we have a Morse function $f: M \to \mathbb{R}$ and an associated gradient-like vector field X. If p_i is a critical point of f, set $c_i = f(p_i)$ and denote the index of p_i by λ_i.

Definition 3.14. Consider the set

$$M_{[c_{i-1}+\varepsilon, c_i+\varepsilon]} = \{x \in M : c_{i-1}+\varepsilon \le f(x) \le c_i+\varepsilon\}.$$

The set of points p in this set that converge to the critical point p_i along integral curves of X as $t \to \infty$ is called the *lower disc* of p_i (we include p_i in this set). Similarly, the set of points p that converge to p_i as $t \to -\infty$ is called the *upper disc.* We denote these discs by $D_l(p_i)$ and $D_u(p_i)$, respectively.

The motivation for this terminology is clear: since we think of the gradient X as pointing "upward" relative to the "height" determined by f, the lower (resp. upper) disc consists of those points lying below (resp. above) p_i. See Figure 3.3 for an illustration.

Remark 3.15. The upper disc $D_u(p)$ is sometimes called the *stable manifold* of the critical point p and the lower disc $D_l(p)$ is called the *unstable manifold.* If we consider the negative gradient of f, points in the stable manifold want to flow toward p while those in the unstable manifold want to flow away from p. See Section 4.2 for more details.

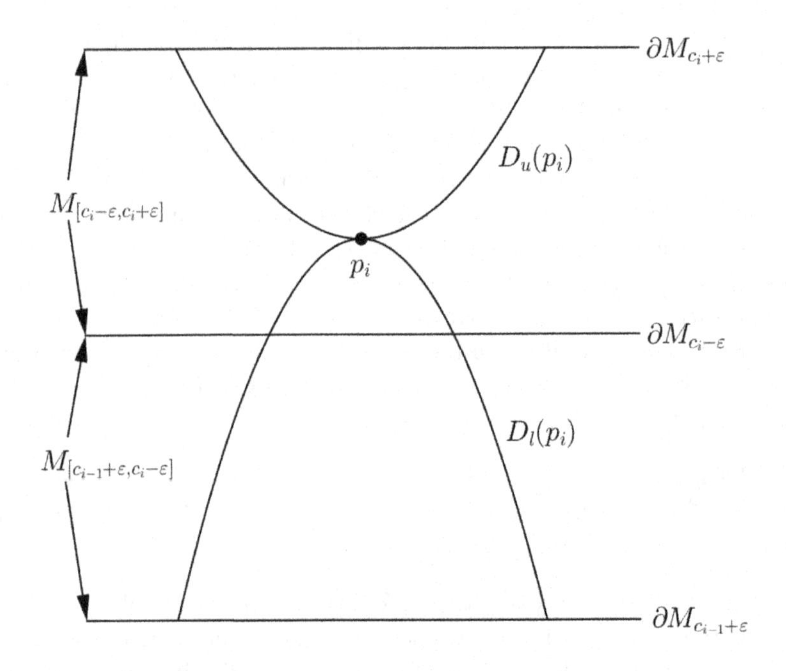

Fig. 3.3 Upper and lower discs near p_i

Since X is the gradient of f written in standard form in a neighborhood of p_i, we see that

$$D_l(p_i) \cap M_{[c_i-\varepsilon,c_i+\varepsilon]} = \{(x_1,\ldots,x_m) : x_1^2 + \cdots + x_{\lambda_i}^2 \leq \varepsilon, x_{\lambda_i+1} = \cdots = x_m = 0\},$$

and hence that this intersection is a disc of dimension λ_i. Moreover, the intersection

$$D_l(p_i) \cap M_{[c_{i-1}+\varepsilon,c_i+\varepsilon]} = \partial(D_l(p_i) \cap M_{[c_i-\varepsilon,c_i+\varepsilon]} \times [c_{i-1}+\varepsilon, c_i - \varepsilon]).$$

Since $D_l(p_i)$ is the union of these two pieces, we see that $D_l(p_i)$ is diffeomorphic to a disc of dimension λ_i.

Similarly, $D_u(p_i)$ may be decomposed as above and we see that it is diffeomorphic to a disc of dimension $m - \lambda_i$. We have met these two discs before in a different guise. The sublevel set $M_{c_i+\varepsilon}$ is obtained from $M_{c_i-\varepsilon} \cong M_{c_{i-1}+\varepsilon}$ by attaching a λ_i-handle. The lower disc $D_l(p_i)$ is the core of the handle and the upper disc $D_u(p_i)$ is the co-core. Observe that the boundary of the lower disc sits inside $\partial M_{c_{i-1}+\varepsilon}$ via the attaching map ψ_i of the λ_i-handle. Furthermore, the boundary of the upper disc $D_u(p_{i-1})$ is a sphere embedded in $\partial M_{c_{i-1}+\varepsilon}$.

Lemma 3.16. *Fix two critical points p_{i-1} and p_i of f. If $\lambda_{i-1} \geq \lambda_i$, then the vector field X may be perturbed to another gradient-like vector field Y so that when replacing the attaching map ψ_i by the new ϕ_i associated to Y we have*

$$\phi_i(\partial D_l(p_i)) \cap \partial D_u(p_{i-1}) = \emptyset.$$

Moreover, we may perform this perturbation while leaving the function f fixed and by not altering the handles attached prior to the ith.

Proof. Note that the dimensions of the discs are

$$\dim D_u(p_{i-1}) = m - \lambda_{i-1} \text{ and } \dim D_l(p_i) = \lambda_i.$$

Then the boundary spheres of these discs satisfy the relation

$$\begin{aligned}\dim \partial D_u(p_{i-1}) + \dim \partial D_l(p_i) &= (m - \lambda_{i-1} - 1) + (\lambda_i - 1)\\ &= m - 2 + (\lambda_i - \lambda_{i-1})\\ &< m - 1.\end{aligned}$$

Since $\dim \partial M_{c_{i-1}+\varepsilon} = m - 1$, Lemma 3.13 implies that there is an isotopy $\{f_t\}_{t\in J}$ of $\partial M_{c_{i-1}+\varepsilon}$ separating $\psi_i(\partial D_l(p_i))$ from $\partial D_u(p_{i-1})$:

$$f_i(\psi_i(\partial D_l(p_i))) \cap \partial D_u(p_{i-1}) = \emptyset.$$

Then by Theorem 3.12 we may replace ψ_i by $\phi_i = f_i \circ \psi_i$ to separate the two boundary spheres. Moreover, we have clearly not altered any attaching maps prior to ψ_i. □

The preceding lemmas show that there is considerable "wiggle room" when it comes to attaching handles. That is, we may move things around in the boundaries of sublevel sets without altering the diffeomorphism type of the manifold. One obvious question arises: do the critical values of the Morse function f really matter? That is, can we perturb the critical values of f without altering the manifold? The following lemma provides the answer.

Lemma 3.17. *Suppose $f : M \to \mathbb{R}$ is a Morse function with associated gradient-like vector field X, and that p_{i-1}, p_i are consecutive critical points ($c_{i-1} < c_i$). Denote by $K(p_{i-1})$ the set of points in $M_{[c_{i-2}+\varepsilon, c_i+\varepsilon]}$ that converge to the critical point p_{i-1} along the integral curves of X as $t \to \pm\infty$, and by $K(p_i)$ the corresponding set for p_i. If $K(p_{i-1}) \cap K(p_i) = \emptyset$ then f may be perturbed to another Morse function g satisfying the following:*

(1) $g = f$ outside $M_{[c_{i-2}+\varepsilon, c_i+\varepsilon]}$;
(2) the sets of critical points and their indices of f and g agree;
(3) for any $a, b \in (c_{i-2}+\varepsilon, c_i+\varepsilon)$, $g(p_{i-1}) = a$ and $g(p_i) = b$.

Proof. The manifold $M_{[c_{i-2}+\varepsilon, c_i+\varepsilon]}$ has boundary consisting of the two disjoint pieces, $\partial_0 = f^{-1}(c_{i-2}+\varepsilon)$ and $\partial_1 = f^{-1}(c_i+\varepsilon)$. An integral curve of X passing through $p \in M_{[c_{i-2}+\varepsilon, c_i+\varepsilon]} \setminus (K(p_{i-1}) \cup K(p_i))$ enters via a point in ∂_0 and leaves through a point in ∂_1. Choose a smooth function h on ∂_0 satisfying $0 \leq h \leq 1$ and having $h \equiv 0$ on a neighborhood of $K(p_{i-1}) \cap \partial_0$ and $h \equiv 1$ on a neighborhood of $K(p_i) \cap \partial_0$. Such a map exists because these intersections are closed and separated in ∂_0. We then define a smooth map $\overline{h} : M_{[c_{i-2}+\varepsilon, c_i+\varepsilon]} \to \mathbb{R}$:

(1) if $p \in M_{[c_{i-2}+\varepsilon, c_i+\varepsilon]} \setminus (K(p_{i-1}) \cup K(p_i))$, set $\overline{h}(p) = h(q)$, where q is the point of ∂_0 where the integral curve of X passing through p hits ∂_0;
(2) $\overline{h} \equiv 0$ on $K(p_{i-1})$;
(3) $\overline{h} \equiv 1$ on $K(p_i)$.

Note that $\overline{h}$ is constant on each integral curve of X (and on $K(p_{i-1})$ and $K(p_i)$). Now let $H : [c_{i-2}+\varepsilon, c_i+\varepsilon] \times [0,1] \to [c_{i-2}+\varepsilon, c_i+\varepsilon]$ be a smooth map satisfying the following:

(1) for a fixed s, $H(x,s)$ strictly increases as a function of x from $c_{i-2}+\varepsilon$ to $c_i+\varepsilon$ as x increases along the interval;
(2) $H(c_{i-1}, 0) = a$ and $H(c_i, 1) = b$;
(3) for a fixed s, $H(x,s) = x$ for x in a sufficiently small neighborhood of $c_{i-2}+\varepsilon$ or $c_i+\varepsilon$;
(4) for x in a neighborhood of c_{i-1},

$$\frac{\partial}{\partial x} H(x,0) = 1;$$

(5) for x in a neighborhood of c_i,

$$\frac{\partial}{\partial x} H(x,1) = 1.$$

We leave the existence of such a function H as an exercise. Now define a function $g : M \to \mathbb{R}$ by setting $g = f$ outside $M_{[c_{i-2}+\varepsilon, c_i+\varepsilon]}$ and on $M_{[c_{i-2}+\varepsilon, c_i+\varepsilon]}$,

$$g(p) = H(f(p), \overline{h}(p)).$$

Then the critical points of g agree with those of f and $g(p_{i-1}) = a$, $g(p_i) = b$. □

We may now prove the result we were looking for.

Theorem 3.18. *Let M be a closed m-manifold and $f : M \to \mathbb{R}$ a Morse function with critical points $p_0, p_1, \ldots, p_n$. Then f may be perturbed to another Morse function g having the same critical points such that if $g(p_i) < g(p_j)$ then* $\mathrm{index}(p_i) \leq \mathrm{index}(p_j)$.

Proof. Let X be a gradient-like vector field for f. We may assume the critical points of f have distinct values; order the p_i so that the corresponding $c_i = f(p_i)$ satisfy

$$c_0 < c_1 < \cdots < c_n.$$

Suppose there is some i such that $\text{index}(p_{i-1}) > \text{index}(p_i)$. By Lemma 3.16 we may alter X to a gradient-like field Y so that $\partial D_l(p_i)$ and $\partial D_u(p_{i-1})$ are disjoint in $f^{-1}(c_{i-1} + \varepsilon)$. Then points in $\partial D_l(p_i)$ do not converge to p_{i-1} when flowing down along Y, and points in $\partial D_u(p_{i-1})$ do not converge to p_i when flowing up along Y. It follows that $K(p_{i-1})$ and $K(p_i)$ are disjoint in $M_{[c_{i-2}+\varepsilon, c_i+\varepsilon]}$. By Lemma 3.17 we may perturb f to a Morse function g satisfying $g(p_{i-1}) > g(p_i)$, while leaving the values at other critical points unchanged. By reindexing the critical points, we therefore fix the problem associated with the pair p_{i-1} and p_i. Repeating this for all such pairs, we obtain the required Morse function g. □

Remark 3.19. Note that a slight modification of the proof of Theorem 3.18 shows that we may attach handles of the same index simultaneously. That is, if $\text{index}(p_{i-1}) = \text{index}(p_i)$, then we may construct the function g to satisfy $g(p_{i-1}) = g(p_i)$. The proof is left as an exercise. In fact, we can prove even more: by rescaling and translating, we may assume that if p is a critical point of index i, then $g(p) = i$. Such a Morse function is called *self-indexing.*

We now turn to the question of minimality. Consider the torus with its standard height function (Figure 1.4). This is a minimal Morse function on the torus in the sense that it has exactly the number of critical points (4) as a minimal cell decomposition for the torus. However, we know that there are many Morse functions on any given manifold and it is unlikely that a particular one will be minimal in this sense.

For example, consider the manifold in Figure 3.4 with the associated height function. This is clearly a Morse function. Moreover, the space is homeomorphic to the torus but the function has 6 critical points. Note, however, that the "bump" on the right is extraneous in the sense that it is easy to see how to deform it away so that it is no longer there. This is the idea behind *handle cancellation.* The critical point p has index 1 and the point q has index 2. In general, we will describe conditions where critical points of indices λ and $\lambda + 1$ can be removed by a perturbation of the Morse function.

Observe the following about the portion of the bumpy torus near the critical points p and q. The upper disc $D_u(p)$ consists of a curve heading uphill toward q while the lower disc $D_l(q)$ is a 2-disc covering the cap surrounding the local maximum at q. For a sufficiently small ε, these discs intersect orthogonally at a single point in the level set $f = f(p) + \varepsilon$. These are exactly the conditions we seek in order to be able to remove such extra critical points.

Definition 3.20. Let N be an n-manifold with submanifolds K and L of dimensions k and ℓ, respectively, where $k + \ell = n$. We say that K and L *inter-*

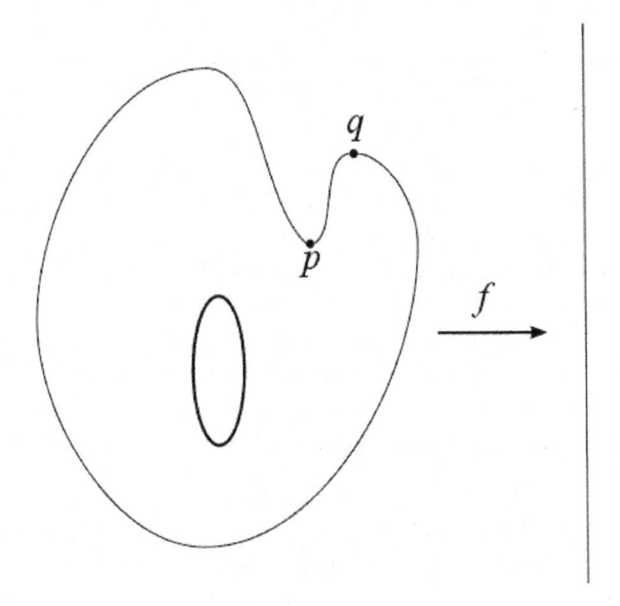

Fig. 3.4 A bumpy torus

sect transversely at a point x of N if there is a coordinate neighborhood U and a local coordinate system $(x_1, \ldots, x_n)$ such that $K \cap U$ consists of the points satisfying $x_{k+1} = x_{k+2} = \cdots = x_n = 0$ and $L \cap U$ consists of points satisfying $x_1 = x_2 = \cdots x_k = 0$. Note that x is the origin in these coordinates.

Thus, we see that the boundary of the upper disc $D_u(p)$ and the boundary of the lower disc $D_l(q)$ intersect transversely (in a single point) in the case of the bumpy torus. The situation for the absolute minimum u of f and the saddle point v at the bottom of the hole in the torus is different, however. The lower disc $D_l(v)$ is a curve centered at v traveling down the torus on either side of the hole toward u. The upper disc $D_u(u)$ is a 2-disc forming the bottom of the torus. These two intersect in a pair of points in any level set $f = f(u) + \varepsilon$. Another way to think of this is that there is a single gradient path between p and q, while there are two such ways to get from u to v.

Theorem 3.21. *Suppose $f : M \to \mathbb{R}$ is a Morse function with critical points $p_0, p_1, \ldots, p_n$. Set $c_i = f(p_i)$ and assume the critical points are indexed so that the c_i form an increasing sequence. Fix an i and suppose the following*

(1) $\text{index}(p_i) = 1 + \text{index}(p_{i-1})$*;*
(2) $\partial D_l(p_i)$ *intersects* $\partial D_u(p_{i-1})$ *transversely at a single point in the level surface* $f^{-1}(c_{i-1} + \varepsilon)$.

Then f may be perturbed to a Morse function g such that $g = f$ near the bound-

ary and outside of $M_{[c_{i-2}+\varepsilon,c_i+\varepsilon]}$ *and* g *has no critical points in the interior of* $M_{[c_{i-2}+\varepsilon,c_i+\varepsilon]}$.

Proof. A complete proof would fill several pages. There are some very technical constructions that, frankly, are not especially illuminating. We shall content ourselves with an outline; a complete proof may be found in [Milnor (1965b)].

The spaces $\partial D_l(p_i)$ and $\partial D_u(p_{i-1})$ intersect transversely at the point $p \in f^{-1}(c_{i-1}+\varepsilon)$. Let X be a gradient-like vector field and let $\sigma = \sigma(t)$ be the integral curve of X passing through p. Note that $\sigma(t) \to p_i$ as $t \to +\infty$ and $\sigma(t) \to p_{i-1}$ as $t \to -\infty$. Moreover, σ is the *only* integral curve of X with this property. In essence, we wish to turn the curve σ around and do so smoothly in a neighborhood of it; this will have the effect of making the points p_i and p_{i-1} into regular points for f. See Figure 3.5 for an illustration.

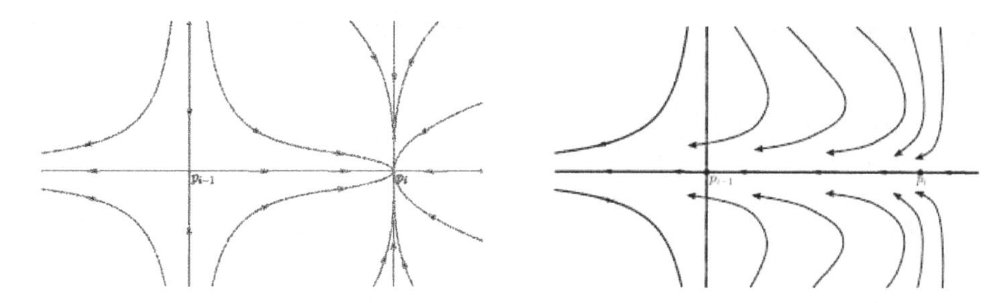

Fig. 3.5 A gradient-like field X and the turned around version Y

Denote by λ the index of p_{i-1}. There is a neighborhood V of p_{i-1} in which X may be written

$$X = 2x_1\frac{\partial}{\partial x_1} - \cdots - 2x_\lambda\frac{\partial}{\partial x_\lambda} - 2x_{\lambda+1}\frac{\partial}{\partial x_{\lambda+1}} + \cdots + 2x_m\frac{\partial}{\partial x_m}.$$

(Here, we have interchanged the role of x_1 and x_λ for reasons that will become clear later.) There is also a neighborhood W of p_i where X has the form

$$X = -2y_1\frac{\partial}{\partial y_1} - \cdots - 2y_{\lambda+1}\frac{\partial}{\partial y_{\lambda+1}} + 2y_{\lambda+2}\frac{\partial}{\partial y_{\lambda+2}} + \cdots + 2y_m\frac{\partial}{\partial y_m}.$$

The technical heart of the proof, available in full detail in Chapter 5 of [Milnor (1965b)] is that there exists a coordinate system $(z_1, \ldots, z_m)$ in a neighborhood U of the curve σ with the following properties.

(1) In these coordinates, $p_{i-1} = (0, 0, \ldots, 0)$ and $p_i = (1, 0, \ldots, 0)$; and
(2) On U, X takes the form

$$X = 2v(z_1)\frac{\partial}{\partial z_1} - \cdots - 2z_{\lambda+1}\frac{\partial}{\partial z_{\lambda+1}} + 2z_{\lambda+2}\frac{\partial}{\partial z_{\lambda+2}} + \cdots + 2z_m\frac{\partial}{\partial z_m}.$$

The existence of U and the coordinate system $(z_1, \dots, z_m)$ depend on the fact that the boundaries of the lower and upper discs of p_i and p_{i-1} intersect transversely in a single point. The function $v(z_1)$ is a smooth function defined on an interval $(-\delta, 1+\delta)$ for some small $\delta > 0$. Moreover, $v'(z_1) = 1$ in a neighborhood of 0, $v'(z_1) = -1$ in a neighborhood of 1, and $v(z_1) > 0$ in the open interval $(0, 1)$. The graph of such a v looks a bit like a parabola opening downward, passing through the origin and the point $(1, 0)$ on the z_1-axis. Assuming all of this, we see that in the coordinates $(z_1, \dots, z_m)$, X has the correct form in neighborhoods of the points p_{i-1} and p_i (for p_i, use the coordinate change $(y_1, \dots, y_m) = (z_1 - 1, z_2, \dots, z_m)$).

It may be that the vector field X vanishes somewhere on U, so we need to adjust it to ensure that $X \neq 0$ everywhere in U. Define an isotopy $F : (-\delta, 1+\delta) \times (-\mu, 2\mu) \to \mathbb{R}$ so that

(1) If $s \geq \mu$, $F(z_1, s) = v(z_1)$;
(2) If $s \leq 0$, $F(z_1, s) = F(z_1, 0)$;
(3) $F(z_1, 0) < 0$ for all z_1;
(4) If $z_1 < -\delta/2$ or $z_1 > 1 + \delta/2$, $F(z_1, s) = v(z_1)$ for all s.

One should visualize a deformation of the graph of $v(z_1)$ which pushes the arch down smoothly so that the graph lies below the z_1-axis once $s = 0$. For each s, denote the function $F(-, s)$ by $v_s(-)$. Now set $\rho : U \to \mathbb{R}$ to be $\rho(z_1, \dots, z_m) = z_2^2 + \cdots + z_m^2$ and define a new vector field $\overline{X}$ on U by

$$\overline{X} = 2v_\rho(z_1)\frac{\partial}{\partial z_1} - 2z_2\frac{\partial}{\partial z_2} - \cdots - 2z_{\lambda+1}\frac{\partial}{\partial z_{\lambda+1}} + \cdots + 2z_m\frac{\partial}{\partial z_m}.$$

By the construction of the $v_s(z_1)$, we see that $\overline{X}$ agrees with X away from the curve σ. Moreover, since the first term of $\overline{X}$ satisfies $2v_0(z_1) < 0$ on the z_1-axis, we see that $\overline{X} \neq 0$ in a neighborhood of σ (which lies along the z_1-axis) and therefore $\overline{X} \neq 0$ everywhere in U. Since $\overline{X}$ agrees with X away from σ, it can be extended smoothly to X outside of U. Denote this extension by Y. Note that all integral curves of Y enter $M_{[c_{i-2}+\varepsilon, c_i+\varepsilon]}$ through the level set $f^{-1}(c_{i-2}+\varepsilon)$ and exit through $f^{-1}(c_i+\varepsilon)$. As we have seen before, we may then define a smooth function $\overline{f} : M_{[c_{i-2}+\varepsilon, c_i+\varepsilon]} \to \mathbb{R}$ which increases along the integral curves of Y and which agrees with f near the boundary. We may therefore extend $\overline{f}$ to the requisite Morse function g on all of M having Y as a gradient-like vector field. This completes the proof. □

In terms of handlebodies, Theorem 3.21 says the following. Suppose we have constructed the handlebody N associated to the Morse function f up through the $(j-1)$th attaching map and that the critical points p_j and p_{j+1} satisfy the conditions of the theorem (i.e., $\text{index}(p_{j+1}) = 1 + \text{index}(p_j)$ and the upper and lower discs intersect transversely in a single point). Let N' be the handlebody obtained from attaching the handle associated to p_j and let N'' be the result of attaching the handle associated to p_{j+1}. Then Theorem 3.21 implies that N'' is diffeomorphic to N; that is, the extra handles that are attached "cancel" each other out.

Here is another interesting application of the preceding results.

Theorem 3.22. *If M is a connected closed m-manifold then there is a Morse function $f : M \to \mathbb{R}$ with exactly one critical point of index 0 and exactly one critical point of index m.*

Proof. Theorem 3.18 and Remark 3.19 imply that there is a Morse function f such that all critical points of index i take the same value c_i. Moreover, we have $c_0 < c_1 < \cdots < c_m$. Then $M_{c_0+\varepsilon}$ is the disjoint union of a collection of m-discs and $M_{c_1+\varepsilon}$ is obtained from this by attaching some 1-handles. If $M_{c_1+\varepsilon}$ is not connected, then M could not be connected. Indeed, since the attaching sphere of a handle of index 2 or higher is connected, it must be attached to a single connected component; thus, if $M_{c_1+\varepsilon}$ had more than one component, so would M. But now if there is more than one m-disc in $M_{c_0+\varepsilon}$, say $D_1^m \cup \cdots \cup D_\ell^m$ with $\ell \geq 2$, then D_ℓ^m is connected to some D_j^m by a 1-handle $D^1 \times D^{m-1}$. Since the attaching sphere of this handle intersects D_j^m in exactly one point, these handles cancel by Theorem 3.21. Iterating this argument, we find a Morse function $g : M \to \mathbb{R}$ with a single critical point of index 0.

Now repeat this argument for the Morse function $-g : M \to \mathbb{R}$. It has the same critical points as g, but the indices are dual: if p is a critical point of index λ for g, it is a critical point of index $m - \lambda$ for $-g$. We may then perturb $-g$ to reduce its number of index 0 critical points to 1; call the resulting function h. Then the function $-h$ has the properties we seek. □

3.4 Exercises

(1) Let $f : M \to \mathbb{R}$ and $g : N \to \mathbb{R}$ be Morse functions on the closed manifolds M and N. Let $a, b > 0$ and define $F : M \times N \to \mathbb{R}$ by $F = (a + f)(b + g)$. Show that F is a Morse function if a and b are sufficiently large. Determine the critical points and their indices in terms of those of f and g. In particular, if $M = S^m$ and $N = S^n$, with f and g the respective height functions, deduce a CW structure for the product $S^m \times S^n$.

(2) Complex projective space, $\mathbb{C}P^m$, is the space of complex lines through the origin in $\mathbb{C}^{m+1}$. It is a manifold of (real) dimension $2m$. Denote the points in $\mathbb{C}P^m$ by $[z_0 : z_1 : \cdots : z_m]$, choose real numbers $a_0 < a_1 < \cdots < a_m$, and define $f : \mathbb{C}P^m \to \mathbb{R}$ by

$$f([z_0 : z_1 : \cdots : z_m]) = \frac{a_0|z_0|^2 + \cdots + a_m|z_m|^2}{|z_0|^2 + \cdots + |z_m|^2}.$$

Show that f is a Morse function on $\mathbb{C}P^m$, determine its critical points, and deduce a CW structure for $\mathbb{C}P^m$.

(3) The special unitary group, $SU(m)$ is the subgroup of $U(m)$ consisting of matrices of determinant 1. It is easy to see that $\dim SU(m) = m^2 - 1$. Show that

the restriction of the Morse function $f : U(m) \to \mathbb{R}$ given in Example 3.11 to $SU(m)$ is a Morse function, provided the c_i are sufficiently large relative to c_1. Deduce a CW structure for $SU(m)$.

(4) Prove that $U(m)$ is diffeomorphic to the product $SU(m) \times S^1$ as a smooth manifold, but that $U(m)$ is not isomorphic to $SU(m) \times S^1$ as a Lie group.

(5) Determine the indices of the critical points of the Morse function $f : SO(3) \to \mathbb{R}$ of Example 3.9. Deduce a CW structure.

(6) Compare the cell structures of $SO(3)$ and $\mathbb{R}P^3$. What do you deduce?

(7) Prove the result asserted in Remark 3.19.

Bibliographic notes

The standard results in this chapter may be found in any text on Morse theory; we have therefore often omitted explicit references as in previous chapters. The examples in Section 3.2 are worked out in greater detail in [Matsumoto (1997)]; moreover our presentation of handle cancellation and sliding follows that of [Matsumoto (1997)] as well.

Chapter 4

Homology

In Chapter 3 we showed that a closed manifold M admits a handle decomposition and that it therefore has the homotopy type of a CW-complex with a cell of dimension i for each critical point of index i for a Morse function f on M. In this chapter we show how to use this information to compute the homology of the manifold M and to deduce the Morse inequalities.

Roughly speaking, the homology groups of a topological space measure the number of "holes" of various dimension in a space. Putting this concept on a rigorous foundation took decades of work in the early days of algebraic topology. One difficulty to be overcome was that there are several potentially different homology theories that one may use. For any topological space one may consider singular homology. Simplicial complexes are amenable to simplicial homology theory; CW-complexes may be attacked via cellular homology. The homology of manifolds may be computed (up to torsion) using differential forms and de Rham cohomology. That all these methods give the same answer was finally proved by Eilenberg and Steenrod. Moreover, homology is a homotopy invariant: if X and Y have the same homotopy type, then $H_\bullet(X) \cong H_\bullet(Y)$.

In this book, we shall use the cellular theory. Indeed, in many ways this theory is the most computationally efficient and since we know, thanks to Theorem 3.6, that any manifold has the homotopy type of a CW-complex we will be able to use Morse theory to aid the computation.

4.1 Cellular homology

Suppose X is a finite CW-complex and denote by n the largest dimension of a cell in X. There are many different equivalent approaches to defining the cellular homology of X, but this is the one we will use. For each i, denote by $C_i(X)$ the free abelian group with basis consisting of the i-cells of X. For any j-cell e^j, we have an attaching map $f : \partial e^j \to X^{(j-1)}$, where $X^{(j-1)}$ is the $(j-1)$-skeleton of

X. Define a map $d_j : C_j(X) \to C_{j-1}(X)$ called the *boundary homomorphism* by

$$d_j(e_k^j) = \sum_{\ell} m_{k\ell} e_\ell^{j-1},$$

where the integers $m_{k\ell}$ are defined as follows. Denote by S_ℓ^{j-1} the $(j-1)$-sphere obtained by taking $X^{(j-1)}$ and collapsing the subcomplex

$$X^{(j-2)} \coprod_{r \neq \ell} e_r^{j-1}$$

to a point. Consider the composition

$$\partial e_k^j = S^{j-1} \xrightarrow{f} X^{(j-1)} \to S_\ell^{j-1}.$$

The integer $m_{k\ell}$ is defined to be the degree of this map.

We then have a sequence of groups and homomorphisms

$$0 \to C_n(X) \xrightarrow{d_n} C_{n-1}(X) \xrightarrow{d_{n-1}} \cdots \xrightarrow{d_2} C_1(X) \xrightarrow{d_1} C_0(X) \longrightarrow 0.$$

Denote this by $C_\bullet(X)$. We call this the *cellular chain complex* of X.

Lemma 4.1. *For each $1 \leq i \leq n$, we have $d_{i-1} \circ d_i = 0$.*

Proof. This is a standard result in homology theory; the proof is omitted. □

Lemma 4.1 implies that $\operatorname{im}(d_i) \subseteq \ker(d_{i-1})$. Denote by $Z_i(X)$ the kernel of d_i and by $B_i(X)$ the image of d_{i+1}.

Definition 4.2. The ith *integral homology group of* X is the abelian group

$$H_i(X; \mathbb{Z}) = Z_i(X)/B_i(X).$$

The ith *Betti number* is the rank of this group (i.e., the number of free summands) and is denoted by $\beta_i(X)$.

Remark 4.3. We could define all of this using an arbitrary commutative ring R of coefficients. That is, instead of taking the free abelian groups generated by the cells of X, we could consider the free R-modules on these sets. The resulting homology groups are then modules over R. In particular, if R is a field, the homology groups are vector spaces over the field.

Example 4.4. Let $n \geq 2$ and consider the sphere S^n. A CW-complex structure on S^n is given by taking a single 0-cell x and attaching an n-cell e^n via the only map $\partial e^n \to x$. The cellular chain complex then has $d_n \equiv 0$ and we see that

$$H_i(S^n; \mathbb{Z}) = \begin{cases} \mathbb{Z} & i = 0, n \\ 0 & \text{otherwise.} \end{cases}$$

The case of S^1 is slightly different. A CW-decomposition is given by a 0-cell x and a 1-cell e with the attaching map $\partial e \to x$ mapping both points of ∂e to x. The cellular chain complex is then

$$0 \longrightarrow C_1(S^1) \xrightarrow{d_1} C_0(X) \longrightarrow 0$$

where the map d_1 is given by

$$d_1(e) = x - x = 0.$$

We therefore see that the homology groups are those above in the case $n = 1$ as well.

Example 4.5. Consider the space $\mathbb{R}P^2$. We have a CW-decomposition consisting of a single 0-cell x, a 1-cell e, and a 2-cell f. The boundary maps are

$$\begin{aligned} d_1(e) &= x - x = 0 \\ d_2(f) &= 2e. \end{aligned}$$

It follows that the chain complex $C_\bullet(\mathbb{R}P^2)$ is

$$0 \longrightarrow \mathbb{Z} \xrightarrow{\times 2} \mathbb{Z} \xrightarrow{0} \mathbb{Z} \longrightarrow 0$$

and that the homology groups are

$$H_i(\mathbb{R}P^2; \mathbb{Z}) = \begin{cases} \mathbb{Z} & i = 0 \\ \mathbb{Z}/2 & i = 1 \\ 0 & i \geq 2. \end{cases}$$

However, if we choose the field $\mathbb{Z}/2$ as the coefficient ring then the cellular chain complex is

$$0 \longrightarrow \mathbb{Z}/2 \xrightarrow{\times 2} \mathbb{Z}/2 \xrightarrow{0} \mathbb{Z}/2 \longrightarrow 0.$$

Of course, in this case multiplication by 2 is the zero map and we obtain the following:

$$H_i(\mathbb{R}P^2; \mathbb{Z}2) = \begin{cases} \mathbb{Z}/2 & i = 0, 1, 2 \\ 0 & i > 2. \end{cases}$$

So we see that the coefficient ring does make a difference in the calculation.

Definition 4.6. For a CW-complex X of dimension n, the *Euler characteristic* is

$$\chi(X) = \sum_{i=0}^{n} (-1)^i \beta_i(X).$$

At first glance, it seems that $\chi(X)$ might depend on the coefficient ring used for computing homology. In the case of S^n it does not, since the boundary maps in the chain complex $C_\bullet(X)$ are always 0, independent of the coefficients. It follows that $\chi(S^n) = 1 + (-1)^n$. Consider $\mathbb{R}P^2$. Integrally, we have

$$\beta_i(\mathbb{R}P^2) = \begin{cases} 1 & i = 0 \\ 0 & i \geq 1, \end{cases}$$

and so $\chi(\mathbb{R}P^2) = 1$. With $\mathbb{Z}/2$-coefficients, we have

$$\beta_i(\mathbb{R}P^2; \mathbb{Z}2) = \begin{cases} 1 & i = 0, 1, 2 \\ 0 & i > 2 \end{cases}$$

so that $\chi(\mathbb{R}P^2) = 1 - 1 + 1 = 1$. This is a general phenomenon. For a proof of the following, see e.g., [Munkres (1984)].

Lemma 4.7. (Euler–Poincaré) *Let X be a finite CW-complex. Then*

$$\chi(X) = \sum_{i=0}^{n} (-1)^i \operatorname{rank} C_i(X) = \sum_{i=0}^{n} (-1)^i \#i\text{-cells of } X.$$

□

As an application of Morse theory, we have the following result.

Theorem 4.8. *Suppose M is a connected, closed, oriented manifold of odd dimension. Then $\chi(M) = 0$.*

Proof. Suppose M is such a manifold of dimension $m = 2k+1$. Let $f : M \to \mathbb{R}$ be a Morse function with $n_i(f)$ critical points of index i, $i = 0, \dots, m$. Then $\chi(M) = \sum (-1)^i n_i(f)$. We also have $n_i(-f) = n_{m-i}(f)$, and $\chi(M) = \sum (-1)^i n_i(-f)$. Putting this together we see that

$$\begin{aligned} \chi(M) &= \sum_{i=0}^{m} (-1)^i n_i(-f) \\ &= \sum_{i=0}^{m} (-1)^i n_{m-i}(f) \\ &= -\sum_{j=0}^{m} (-1)^j n_j(f) \\ &= -\chi(M). \end{aligned}$$

It follows that $\chi(M) = 0$. □

4.2 The Morse complex

In the preceding section we used the fact that a Morse function gives rise to a CW-decomposition of a manifold to compute the cellular homology of M. There is another way to view this calculation, however, using the *Morse complex*. We begin by examining how the upper and lower discs of the critical points of a Morse function partition the manifold M.

For $i = 0, \dots, m$, let $\mathbb{M}_i$ be the free abelian group with a basis element for each critical point of f of index i. We want to define a boundary map $d_i : \mathbb{M}_i \to \mathbb{M}_{i-1}$ so that $\mathbb{M}_\bullet$ is a chain complex whose homology agrees with that of M. We therefore must find a way to relate critical points of index i to those of index $i-1$.

To this end, we will use the *negative gradient* of f. The *flow* associated to this is a smooth function $\varphi : \mathbb{R} \times M \to M$ satisfying the differential equation

$$\frac{\partial}{\partial t}\varphi(t,x) = -\nabla f(\varphi(t,x)); \varphi(0,x) = x.$$

For a fixed $x \in M$, the curve $\varphi(t,x)$ is the flow line of the vector field $X = -\nabla f$ passing through x at time $t = 0$. If we think of f as a height function on M, then the flow lines of X run "downhill." With φ in hand, we are able to recast the upper and lower discs of the critical points of f in the following terms.

Definition 4.9. Let p be a critical point of the Morse function f and let φ be the flow associated to the negative gradient $-\nabla f$. The *stable manifold* associated to p is the set

$$W_p^s = \{x \in M : \lim_{t\to\infty} \varphi(t,x) = p\}.$$

The *unstable manifold* associated to p is the set

$$W_p^u = \{x \in M : \lim_{t\to-\infty} \varphi(t,x) = p\}.$$

Note that W_p^u is a disc of dimension equal to index(p) and W_p^s is a disc of dimension $m -$ index(p), as we may identify them with the lower and upper discs at p, respectively. Note that the collection of unstable manifolds is a cell decomposition of the manifold M, but in general the cells may not intersect nicely enough to use this to compute homology effectively.

Definition 4.10. The Morse function $f : M \to \mathbb{R}$ satisfies the *Morse–Smale* condition if its stable and unstable manifolds intersect transversely.

While there are Morse functions that are not Morse–Smale, this is not the generic case. That is, given any Morse function there is an arbitrarily small perturbation of it that is Morse–Smale (more accurately, all of this involves the choice of a Riemannian metric on M (see Appendix A), and the transversality condition is generic for this choice). Moving forward, we will assume that our function f is actually Morse–Smale without further mention.

Definition 4.11. If p and q are critical points of f, a *flow line* is a curve $\gamma : \mathbb{R} \to M$ such that $\gamma'(t) = -\nabla f(\gamma(t))$, $\lim_{t\to-\infty} \gamma(t) = p$, and $\lim_{t\to\infty} \gamma(t) = q$.

Now, if we intersect the unstable and stable manifolds of f, we get a cell decomposition of M, called the *Morse–Smale complex*. It is a CW-complex structure on M; this is guaranteed by the transversality condition. The vertices are the critical points of f, and the cells consist of the union of flow lines joining the critical points. Observe that the set of flow lines from p to q is the intersection

$$W_p^u \cap W_q^s,$$

and this cell has dimension

$$\begin{aligned}\dim(W_p^u \cap W_q^s) &= \dim W_p^u + \dim W_q^s - m \\ &= \mathrm{index}(p) + (m - \mathrm{index}(q)) - m \\ &= \mathrm{index}(p) - \mathrm{index}(q).\end{aligned}$$

There is an action of the additive group $\mathbb{R}$ on this cell given by $\alpha \cdot \gamma(t) = \gamma(t+\alpha)$. Denote the quotient of the cell by this action by $\mathcal{C}(p,q)$. This is called the moduli space of trajectories. Then if $\mathrm{index}(p) - \mathrm{index}(q) = 1$, this space is 0-dimensional. We claim that it is also compact so that it is in fact a finite set. Moreover, we will put an orientation on $\mathcal{C}(p,q)$ that will allow us to define the boundary map we seek for the Morse complex.

For each critical point p, choose an orientation of the unstable manifold W_p^u. If γ is a flow line from p to q, then we have at any point along γ

$$\begin{aligned}TW_p^u &\cong T(W_p^u \cap W_q^s) \oplus TM/TW_q^s \\ &\cong T_\gamma\mathcal{C}(p,q) \oplus T\gamma \oplus T_q(W_q^u).\end{aligned}$$

Here, the first isomorphism follows from the transversality condition. The second isomorphism is obtained by noting that $\dim \mathcal{C}(p,q) = \mathrm{index}(p) - \mathrm{index}(q) - 1$ (so that $T(W_p^u \cap W_q^s) \cong T_\gamma\mathcal{C}(p,q) \oplus T\gamma$) and that $TM/TW_q^s \cong T_q(W_q^u)$ by translating the subspace $T_q(W_q^u) \subset T_qM$ along γ, keeping it complementary to TW_q^s. This induces an orientation on $\mathcal{C}(p,q)$ if we insist that this isomorphism be orientation preserving.

Now, in the case we are most interested in, namely $\mathrm{index}(p) - \mathrm{index}(q) = 1$, we have the following result. A *smooth manifold with corners* is a second countable Hausdorff space where each point has a neighborhood homeomorphic to $\mathbb{R}^{n-k} \times [0,\infty)^k$ for some k and the transition maps between neighborhoods are smooth. The main result we need to define the boundary map is the following ([Schwarz (1993)]).

Theorem 4.12. *If M is closed and f is Morse–Smale, then for any two critical points p and q, the space $\mathcal{C}(p,q)$ admits a natural compactification to a stratified smooth manifold with corners $\overline{\mathcal{C}(p,q)}$ with codimension k stratum*

$$\overline{\mathcal{C}(p,q)}_k = \bigcup_{r_1,\ldots,r_k} \mathcal{C}(p,r_1) \times \mathcal{C}(r_1,r_2) \times \cdots \times \mathcal{C}(r_{k-1},r_k) \times \mathcal{C}(r_k,q),$$

where the r_i are distinct from each other and from p and q. When $k=1$, then there is an equality of oriented manifolds

$$\partial\overline{\mathcal{C}(p,q)} = \bigcup_{r \neq p,q} (-1)^{\mathrm{index}(p)+\mathrm{index}(q)+1}\mathcal{C}(p,r) \times \mathcal{C}(r,q).$$

The decomposition in the theorem says that a flow line from p to q may be split into k-pieces in a certain sense.

In particular, if $\text{index}(p) = i$ and $\text{index}(q) = i-1$, then $\mathcal{C}(p,q)$ is compact, as we asserted. If $\text{index}(q) = i-2$, then $\mathcal{C}(p,q)$ has a compactification which is a compact 1-manifold with boundary

$$\partial\overline{\mathcal{C}(p,q)} = \bigcup_{r\in\text{crit}_{i-1}(f)} \mathcal{C}(p,r)\times\mathcal{C}(r,q).$$

We are now ready to define the boundary map in $\mathbb{M}_\bullet$:

$$d_i(p) = \sum_{q\in\text{crit}_{i-1}(f)} \#\mathcal{C}(p,q)q,$$

where $\#\mathcal{C}(p,q)$ is defined as follows. Each point in $\mathcal{C}(p,q)$ has an orientation induced from that in W_p^u. Attach a sign of $+1$ if it agrees with that in W_p^u and -1 otherwise. Then $\#\mathcal{C}(p,q)$ is the sum of these signs.

Remark 4.13. If we are willing to work with $\mathbb{Z}_2$ coefficients for homology, then we can ignore the orientations and simply count the points in $\mathcal{C}(p,q)$ modulo 2.

Lemma 4.14. *With the maps d_i defined above, $\mathbb{M}_\bullet$ is a chain complex; that is, $d_{i-1}\circ d_i = 0$ for all i.*

Proof. This follows from the decomposition of $\mathcal{C}(p,q)$: the boundary of the 2-cell determined by two critical points whose indices differ by 2 is an oriented compact 1-manifold and such an object has zero boundary points counted with sign. More accurately, if $x\in\mathbb{M}_\ell$ and y is a basis element of $\mathbb{M}_\ell$ denote by $\langle x,y\rangle$ the coefficient of y in the expression of x with respect to the basis; we then have the following calculation for p a critical point of index i and q a critical point of index $i-2$:

$$\begin{aligned}
\langle d_{i-1}(d_i(p)),q\rangle &= \sum_{r\in\text{crit}_{i-1}(f)} \langle d_i(p),r\rangle\langle d_{i-1}(r),q\rangle \\
&= \#\bigcup_{r\in\text{crit}_{i-1}(f)} \mathcal{C}(p,r)\times\mathcal{C}(r,q) \\
&= \#\partial\overline{\mathcal{C}(p,q)} \\
&= 0.
\end{aligned}$$

□

Definition 4.15. The *Morse homology* $H_\bullet(f)$ is the homology of the chain complex $\mathbb{M}_\bullet$.

Remark 4.16. Strictly speaking, this definition depends on a choice of Riemannian metric (which we have suppressed) and the function f.

Example 4.17. Consider the bumpy torus shown in Figure 3.4. The height function shown is not Morse–Smale since there are two flow lines between the saddle at the top of the hole and the saddle at the bottom (Morse–Smale functions do not have flow lines between critical points of the same index). However, we may perturb the metric slightly to eliminate these, leaving the other flow lines unaltered. We then have 6 critical points: 1 minimum, 2 maxima, and 3 saddles. Consider the 4 critical points different from p and q. There are two flow lines from the maximum m to each of the saddles and they have opposite orientation; it follows that $d_2(m) = 0$. Similarly, there are two flow lines from each of the saddles r_1, r_2 to the minimum with opposite orientation and so $d_1(r_1) = d_1(r_2) = 0$. Now consider the pair q, p. There is exactly one flow line from q to p and no flow lines from q to the other saddles r_1, r_2; thus $d_1(q) = \pm p$, the sign depending on the choices of orientations. In any case, this is unimportant and we see that

$$H_i(f) = \begin{cases} \mathbb{Z} & i = 0, 2 \\ \mathbb{Z}^2 & i = 1 \\ 0 & \text{otherwise.} \end{cases}$$

Note that this agrees with the ordinary homology of the torus.

Theorem 4.18. *If M is a closed smooth manifold and f is a Morse–Smale function on M, then there is a canonical isomorphism*

$$H_\bullet(f) \cong H_\bullet(M).$$

Proof. As we have seen many times, the idea of the proof is fairly simple but there are many technicalities. Given a critical point p of index k, we define a map $\mathbb{M}_k \to C_k(M)$ by sending p to its unstable manifold W_p^u, viewed as a chain. A map in the other direction is obtained by taking a cell and letting it flow along the gradient of f. One composition is the identity on the chain level and the other is chain homotopic to the identity; the result follows.

The classical approach, as presented in Section 7 of [Milnor (1965b)] is the following. Assuming that Morse homology is independent of the choices of f and a metric g (which may be proved directly and which should be unsurprising), we may as well assume that f is *self-indexing*: if p is a critical point of index i, then $f(p) = i$. Then there is an obvious isomorphism

$$\Gamma_i : \mathbb{M}_i \overset{\cong}{\to} H_i^{\text{sing}}(M_i, M_{i-1}; \mathbb{Z}),$$

where M_i is the sublevel set $f^{-1}(-\infty, i]$, and H_i^{sing} denotes singular homology. The main issue is to show that these maps commute with the boundary operators in the two chain complexes:

$$\Gamma_{i-1} \circ d_i = \partial_i \circ \Gamma_i,$$

where ∂_i is the boundary operator arising from the long exact sequence of the triple (M_i, M_{i-1}, M_{i-2}):

$$H_i(M_i, M_{i-1}) \overset{\partial_i}{\to} H_{i-1}(M_{i-1}, M_{i-2}).$$

We omit the details, referring the reader to [Milnor (1965b)] or [Schwarz (1999)]. □

In Part II, we will construct chain complexes associated to a discrete Morse function on M. The analogous theorems are simpler to prove in that context and will be presented in greater detail.

4.3 The Morse inequalities

Denote by $c_i(X)$ the number of i-cells in the CW-complex X. Then we clearly have

$$c_i(X) \geq \beta_i(X).$$

We therefore have the following result.

Theorem 4.19. (Weak Morse Inequalities) *Suppose $f : M \to \mathbb{R}$ is a Morse function on the smooth compact manifold M. Denote by n_i the number of critical points of f of index i. Then for $i = 0, 1, \ldots, m$ we have*

$$n_i \geq \beta_i(M).$$

Proof. M has a CW-decomposition with an i-cell for each critical point of f of index i. □

More is true, however. Note that since M has a CW-decomposition with n_i cells of index i, we have

$$\chi(M) = \sum_{i=0}^{m} (-1)^i n_i.$$

From this we deduce the following set of inequalities.

Theorem 4.20. (Strong Morse Inequalities) *For each $i = 0, \ldots, m$,*

$$n_i - n_{i-1} + \cdots + (-1)^i n_0 \geq \beta_i(M) - \beta_{i-1}(M) + \cdots + (-1)^i \beta_0(M).$$

Proof. Define the Morse polynomial of M associated to f to be

$$\mathcal{M}_f(t) = \sum_{i=0}^{m} n_i t^i,$$

and the Poincaré polynomial to be

$$\mathcal{P}(t) = \sum_{i=0}^{m} \beta_i.$$

We claim there is a polynomial $Q(t)$ with nonnegative integer coefficients such that

$$\mathcal{M}_f(t) - \mathcal{P}(t) = (1 + t)Q(t).$$

For a real number a, denote by $\mathcal{M}_f^a(t)$ the Morse polynomial for the sublevel set $M_a = f^{-1}(-\infty, a]$, and by $\mathcal{P}^a(t)$ the Poincaré polynomial for M_a. Say c_0 is the minimum of the function f. Then we clearly have $\mathcal{M}_f^{c_0}(t) = \mathcal{P}^{c_0}(t)$ (with both equal to the constant polynomial giving the number of points in $f^{-1}(c_0)$). If there are no critical values in an interval $[a, b]$, then $\mathcal{M}_f^a(t) = \mathcal{M}_f^b(t)$, and by Theorem 3.1, the Poincaré polynomials agree as well. Now suppose there is a single critical value in the interval (a, b) with a critical point of index λ. Then

$$\mathcal{M}_f^b(t) - \mathcal{M}_f^a(t) = t^\lambda.$$

The Poincaré polynomial changes depending on whether the λ-handle that gets attached bounds an existing $(\lambda-1)$-cycle or creates a new λ-cycle. In the former case we have

$$\mathcal{P}^b(t) - \mathcal{P}^a(t) = -t^{\lambda-1},$$

and therefore the difference is

$$\mathcal{M}_f^b(t) - \mathcal{P}^b(t) = t^\lambda + t^{\lambda-1} = (1+t)t^{\lambda-1}.$$

In the latter case, we have

$$\mathcal{P}^b(t) - \mathcal{P}^a(t) = t^\lambda,$$

and therefore the difference is

$$\mathcal{M}_f^b(t) - \mathcal{P}^b(t) = t^\lambda - t^\lambda = 0.$$

This completes the inductive step. The result follows by taking b large enough so that $M_b = M$. □

Definition 4.21. A Morse function f is *perfect* if $\mathcal{M}_f(t) = \mathcal{P}(t)$.

Example 4.22. The height function on the sphere S^n is perfect for any $n \geq 1$. Indeed, this function has two critical points, one of index 0 and one of index n, and these match the Betti numbers of S^n. Similarly, the standard height function on the torus is perfect.

Proposition 4.23. *Let f be a Morse function on the closed, connected manifold M. Suppose that for any pair of critical points p, q of f, $|\text{index}(p) - \text{index}(q)| \neq 1$. Then f is a perfect Morse function.*

Proof. If a critical point of index i exists, then f has no critical points of index $i-1$ or $i+1$. Thus, if $C_i(X) \neq 0$, we must have $C_{i+1}(X) = 0$ and $C_{i-1}(X) = 0$. It follows that all maps in the chain complex $C_\bullet(X)$ must vanish and so

$$n_i = \dim C_i(X) = H_i(X) = \beta_i$$

for all i. □

4.4 Poincaré duality

We have seen (Theorem 3.22) that any m-manifold supports a Morse function with a single critical point of index 0 and a single critical point of index m. This was proved via a handle cancellation argument to eliminate all but one index-0 point and then repeating the argument on the negative function to remove all but one index-m point. This is a sort of duality result which we now generalize. To this end, we need to define the cohomology groups of a manifold.

If X is a CW complex, denote by $C_\bullet(X)$ the cellular chain complex. The *dual cochain complex* $C^\bullet(X)$ is obtained by applying the functor $\mathrm{Hom}_\mathbb{Z}(-,\mathbb{Z})$:

$$C^\bullet(X) = \mathrm{Hom}_\mathbb{Z}(C_\bullet(X),\mathbb{Z}) = \{0 \to C^0(X) \overset{\delta^0}{\to} C^1(X) \overset{\delta^1}{\to} \cdots \overset{\delta^{m-1}}{\to} C^m(X) \to 0\},$$

where each $C^i(X) = \mathrm{Hom}_\mathbb{Z}(C_i(X),\mathbb{Z})$. Cochains are therefore integer-valued functions on the cellular chains of X. We define the *i-cocycles* to be $Z^i = \ker \delta^i$, the *i-coboundaries* to be $B^i = \mathrm{im}\, \delta^{i-1}$, and the *$i$th cohomology group* to be

$$H^i(X,\mathbb{Z}) = Z^i/B^i.$$

If $f \in Z^i(X)$, denote its cohomology class by $[f]$; similarly, if $c \in Z_i(X)$, denote its homology class by $[c]$.

Lemma 4.24. *The value $f(c)$ depends only on the cohomology class $[f]$ and homology class $[c]$.*

Proof. Any element of $[f]$ has the form $f + \delta^{i-1}g$ for some $g \in C^{i-1}(X)$ and any element of $[c]$ has the form $c + d_{i+1}s$ for some $s \in C_{i+1}(X)$. Then

$$\begin{aligned}(f + \delta^{i-1}g)(c + d_{i+1}s) &= f(c) + f(d_{i+1}(s)) + \delta^{i-1}g(c) + \delta^{i-1}g(d_{i+1}(s)) \\ &= f(c) + \delta^i f(s) + g(d_i(c)) + \delta^i\delta^{i-1}g(s) \\ &= f(c) + 0 + 0 + 0 \\ &= f(c).\end{aligned}$$

□

We therefore have a well-defined homomorphism $[f] : H_i(X) \to \mathbb{Z}$ given by $[f][c] = f(c)$. Define a map $\varphi : H^i(X) \to \mathrm{Hom}(H_i(X),\mathbb{Z})$ by $\varphi[f][c] = f(c)$. Then we have the Universal Coefficient Theorem (see, e.g., [Munkres (1984)]): the map φ is a surjective homomorphism with kernel equal to the torsion part of $H^i(X)$. (We actually have more: the map φ splits so that $H^i(X)$ contains $\mathrm{Hom}(H_i(X),\mathbb{Z})$ as a direct summand.)

Now, if M is an orientable closed manifold, we have seen that $H_0(M) = \mathbb{Z}$. It is easy to compute the top homology group as well: $H_m(M) = \mathbb{Z}$. This should not be surprising in light of the fact that there is a Morse function on M with a single critical point of index m; that it generates a homology class is not too difficult to see. We then deduce that $H^m(M)$ contains a copy of $\mathbb{Z} = \mathrm{Hom}(H_m(M),\mathbb{Z})$. More is true, however: this is all there is in $H^m(M)$; that is, $H^m(M)$ is torsion-free. Since $H^0(M) = \mathbb{Z}$, we see that $H^m(M) \cong H_0(M)$ and $H^0(M) \cong H_m(M)$. The main result of this section generalizes this to all (co)homology groups.

Theorem 4.25. (Poincaré duality) *If M is an orientable closed manifold, then for all $i = 0, 1, \ldots, m$, we have natural isomorphisms*

$$H^i(M) \cong H_{m-i}(M).$$

Remark 4.26. Knowing what we know about Morse functions, this theorem should not be particularly surprising. The proof hinges on the following simple observation. If f is a Morse function then we have a CW-decomposition of M determined by the critical points of f; turning things upside down using $-f$, we obtain another "dual" CW-decomposition of M where the dimensions of the cells are dual to those for f. We just need to define the right map between i-cochains in the first decomposition and $(m-i)$-chains in the second.

Proof. We first note the following linear algebra fact: given a basis of $C_i(X)$ we have the corresponding *dual basis* of $C^i(X)$: for a basis element $\sigma \in C_i(X)$, the dual basis element in $C^i(X)$ is the characteristic function χ_σ of σ. With respect to the given bases, the matrix of δ^i in terms of the corresponding dual bases is the transpose of $d_i : C_i(X) \to C_{i-1}(X)$.

Now, given a Morse function f on M, we obtain a handlebody decomposition of M:

$$M = (h_1^0 \cup \cdots \cup h_{\ell_0}^0) \cup (h_1^1 \cup \cdots \cup h_{\ell_1}^1) \cup \cdots \cup (h_1^m \cup \cdots \cup h_{\ell_m}^m)$$

where each h_j^i is an i-handle. The associated CW-decomposition of M is obtained by taking the core e_j^i of each i-handle as an i-cell; the attaching maps of the handle decomposition are the attaching maps of the CW-decomposition. Denote the co-core of h_j^i by ϵ_j^{m-i}. To avoid confusion, denote this CW-complex by X. Of course the (co)homology of X is canonically isomorphic with that of M.

Fix i and let N^i be the subhandlebody obtained by attaching all the handles up to and including the i-handles. Denote by S_j^i the attaching sphere of h_j^{i+1}; $S_j^i = \psi_j(\partial D^{i+1} \times 0)$, where ψ_j is the attaching map of the $(i+1)$-handle. Denote by Σ_k^{m-i-1} the boundary sphere of the co-core of h_k^i; $\Sigma_k^{m-i-1} = 0 \times \partial D^{m-i}$. Σ_k^i is sometimes called the *belt sphere* of the handle. Note that the dimensions of the attaching sphere and the belt sphere add up to the dimension $m-1$ of ∂N^i.

Recall the general position statement of Lemma 3.13. We need a slight generalization of this (see, e.g., [Milnor (1965b)]). Suppose we have two closed submanifolds A and B in a smooth manifold K with $\dim A + \dim B = \dim K$. Then there is an isotopy $\{h_t\}_{t \in J}$ such that (i) $h_0 = \mathrm{id}_K$, and (ii) $h_1(A)$ intersects B transversely in finitely many points. Now, since we can slide handles at will (Theorem 3.12), we may as well assume that the attaching sphere S_j^i intersects the belt sphere Σ_k^{m-i-1} transversely in finitely many points.

Suppose A^a and B^b are orientable submanifolds of K $(a + b = \dim K)$ which intersect transversely at finitely many points. If q is an intersection point choose an a-frame of vector fields $\langle V_1, \ldots, V_a \rangle$ on A compatible with the chosen orientation and a b-frame $\langle W_1, \ldots, W_b \rangle$ on B in some neighborhood of q. (See Appendix A for the relevant definitions.) The *sign of intersection* $\sigma(q)$ is defined to be ± 1 depending on whether $\langle V_1, \ldots, V_a, W_1, \ldots, W_b\rangle$ matches the orientation of M.

Definition 4.27. If $A \cap B = \{q_1, \ldots, q_r\}$, the *intersection number* of A and B is

$$\langle A\rangle \cdot \langle B\rangle = \sum_{i=1}^{r} \sigma(q_i).$$

Note the following fact, whose proof is left as an exercise:

$$\langle A\rangle \cdot \langle B\rangle = (-1)^{ab}\langle B\rangle \cdot \langle A\rangle.$$

Now we claim the following about the boundary map in the chain complex $C_\bullet(X)$. If a_{jk} denotes the coefficient of $d_{i+1}(e_j^{i+1})$ associated with the i-cell e_k^i, we assert that

$$a_{jk} = (-1)^i \langle S_j^i\rangle \cdot \langle \Sigma_k^{m-i-1}\rangle.$$

We know that this number is the number of times that the attaching sphere S_j^i covers the cell e_k^i, but this is easily seen to be the same as the number of times S_j^i intersects Σ_k^{m-i-1} (taking signs into account). The trick is to choose the orientations correctly so that we get the claimed formula. To this end, orient the core e_k^i arbitrarily. Since the core and co-core ϵ_k^{m-i} intersect in a single point, the orientation of e_k^i determines an orientation on the co-core so that $\langle e_k^i\rangle \cdot \langle \epsilon_k^{m-i}\rangle = 1$ (given a global orientation for M). The orientation on ∂N^i is the one induced from the orientation on N^i. With these choices, the formula holds.

Consider the dual decomposition of M obtained via the Morse function $-f$. In this decomposition, the cores and co-cores reverse roles. The dual decomposition is

$$M = (k_1^0 \cup \cdots \cup k_{\ell_m}^0) \cup (k_1^1 \cup \cdots \cup k_{\ell_{m-1}}^1) \cup \cdots \cup (k_1^m \cup \cdots \cup k_{\ell_0}^m),$$

where the core of k_j^{m-i} is ϵ_j^{m-i} and the co-core is e_k^i. Denote this CW-decomposition of M by X^*.

The basis of $C_{m-i}(X^*)$ consists of $\langle \epsilon_1^{m-i}\rangle, \cdots, \langle \epsilon_{\ell_i}^{m-i}\rangle$. Note that this has the same number of elements as the dual basis of $C^i(X)$, using intersection numbers. Indeed, we have the following for $\langle \epsilon_k^{m-i}\rangle \in C_{m-i}(X^*)$ and $\langle e_j^i\rangle \in C_i(X)$:

$$\langle \epsilon_k^{m-i}\rangle \cdot \langle e_j^i\rangle = \begin{cases} (-1)^{i(m-i)} & j = k \\ 0 & j = k. \end{cases}$$

Note also that the groups $C_{m-i}(X^*)$ and $C^i(X)$ have the same rank, equal to ℓ_i. Define an isomorphism $\Phi_{m-i} : C_{m-i}(X^*) \to C^i(X)$ by

$$\Phi(\langle \epsilon_k^{m-i}\rangle) = (-1)^{(m-i)i}\chi_{e_k^i}.$$

Consider the following diagram

$$\begin{array}{ccc} C^i(X) & \xrightarrow{\delta^i} & C^{i+1}(X) \\ \uparrow{\scriptstyle \Phi_{m-i}} & & \uparrow{\scriptstyle \Phi_{m-i-1}} \\ C_{m-i}(X^*) & \xrightarrow[d^*_{m-i}]{} & C_{m-i-1}(X^*) \end{array}$$

Claim. This diagram commutes up to sign: $\Phi_{m-i-1} \circ d^*_{m-i} = \pm\delta^i \circ \Phi_{m-i}$.

This statement amounts to the following assertion: the matrices for δ^i and d^*_{m-i} agree up to sign. That is, we must show that the matrix for d^*_{m-i} is the transpose of that for d_i, up to sign. But this is more or less clear given that the bases for $C_{m-i}(X^*)$ and $C_{m_i-1}(X^*)$ are dual to those for $C_i(X)$ and $C_{i+1}(X)$: the core of an $(m-i)$-handle in X^* is the co-core of an i-handle in X and the index of the attaching map may be obtained by counting intersection points of the attaching spheres and belt spheres. The only issue is that of induced orientations. If we denote by K^{m-i} the handlebody decomposition obtained from $-f$ by attaching all the handles up to and including the $(m-i)$-handles, then $\partial K^{m-i} = \partial N^i$ but the orientations are opposite. This will affect the intersection numbers of the spheres, but only by a sign. The details are left as an exercise.

Now, given this, we see that the chain complex $C_\bullet(X^*)$ is isomorphic to the cochain complex $C^\bullet(X)$ and therefore we have an isomorphism (induced by Φ_{m-i}) for each i:

$$H_{m-i}(X^*) \cong H^i(X).$$

Since these groups are canonically isomorphic to the (co)homology groups of M, this completes the proof. □

Corollary 4.28. *If M is an orientable, connected, closed m-manifold, then for all $i = 0, 1, \ldots, m$*

$$\beta_i(M) = \beta_{m-i}(M).$$

Proof. The universal coefficient theorem implies that the rank of $H^i(M)$ agrees with the rank of $H_i(M)$. Poincaré duality then implies the result. □

4.5 Exercises

(1) Let $f : S^1 \to \mathbb{R}$ be a Morse function. Show that f has an even number of critical points, half of which are local minima.

(2) Consider the Morse function $f : S^1 \times S^1 \to \mathbb{R}$ defined by

$$f(\theta, \phi) = (R + r\cos\phi)\cos\theta.$$

Construct the associated Morse complex $H_\bullet(f)$ and show that it computes the homology of the torus.

(3) Prove that

$$\langle A\rangle \cdot \langle B\rangle = (-1)^{ab}\langle B\rangle \cdot \langle A\rangle.$$

(4) Prove that the diagram in the proof of Theorem 4.25 commutes up to sign by showing that the boundary matrix d^*_{m-i} for the dual decomposition X^* satisfies $(-1)^{i+1}d_i$.

Bibliographic notes

As in the previous chapters, theorems which are not referenced explicitly may be found in standard texts on Morse theory. Our proof of Poincaré Duality follows that of Matsumoto [Matsumoto (1997)]. Full details of Morse homology fill an entire book on their own; we heartily recommend the book of Schwarz [Schwarz (1999)] to the interested reader.

Chapter 5

Piecewise Linear Morse Theory

In previous chapters we developed the fundamentals of Morse theory on smooth manifolds, but it should be clear that it would be useful to discuss these concepts in more general settings. For example, any smooth manifold admits a triangulation and a natural class of functions to consider on such spaces are *piecewise linear* functions; that is, maps f which are linear on each simplex in the sense that if we express a point in barycentric coordinates, $\sum b_i v_i$, then

$$f\left(\sum b_i v_i\right) = \sum b_i f(v_i).$$

While easy to write down, such maps have many problems. More generally, one might consider polyhedra embedded in Euclidean space and attempt to understand their topology via some version of Morse theory; for example, is there some way to define critical points relative to a "height function" so that some sum of the indices adds up to the Euler characteristic? In this chapter we examine these questions.

5.1 Critical points on embedded polyhedral surfaces

T. Banchoff [Banchoff (1970)] developed a theory of critical points for polyhedral surfaces embedded in $\mathbb{R}^3$, and proved that it captures the topology of such surfaces. Indeed, any closed orientable surface S of genus g is determined up to homeomorphism by its Euler characteristic $\chi(S) = 2-2g$. In this section we present Banchoff's basic results in this direction. It turns out that his idea is the impetus for much of the discrete Morse theory that follows, beginning with the computational Morse theory of Edelsbrunner, Harer, and Zomorodian in Section 5.2 and continuing on to Algorithm 8.22.

Suppose S is a closed smooth surface embedded in $\mathbb{R}^3$. If we have a Morse function $f : S \to \mathbb{R}$ then we may classify the critical points of f by their indices (0 for minima, 1 for saddles, and 2 for maxima), and then deduce that S is homotopy equivalent to a cell complex with the corresponding numbers of 0-, 1-, and 2-cells. Note, however, that since the Euler characteristic of S may be computed either as the alternating sum of the Betti numbers or the alternating sum of the numbers of critical points of various indices, we have

$$\chi(S) = \#(\text{local minima}) - \#(\text{saddles}) + \#(\text{local maxima}).$$

Now, suppose the function f is a linear function $\pi_{\mathbf{v}}$ given by projecting $\mathbb{R}^3$ onto the line determined by a unit vector $\mathbf{v}$ (think of the standard height function on a sphere or torus, for example). Then p will be a critical point for $\pi_{\mathbf{v}}$ precisely when the tangent plane to S at p is orthogonal to $\mathbf{v}$. For generic choices of $\mathbf{v}$ there will be only finitely many such critical points ("generic" here means except on a set of measure zero). Because we wish to extend the idea of a critical point to a polyhedral surface, it will be convenient to redefine the notion of index of a critical point as follows.

Definition 5.1. Let $\pi_{\mathbf{v}}$ be the linear projection of $\mathbb{R}^3$ onto the line determined by the unit vector $\mathbf{v}$ and let p be a critical point for $\pi_{\mathbf{v}}$. The *index* of p is $i(p, \mathbf{v}) = 1$ if p is a local maximum or minimum and $i(p, \mathbf{v}) = -1$ if p is a saddle point.

In light of the remarks above, we see that we have the following equality:

$$\sum_{p \text{ critical}} i(p, \mathbf{v}) = \chi(S).$$

This notion of index has a nice geometric interpretation. Suppose p is a local maximum or minimum for $\pi_{\mathbf{v}}$. Then the tangent plane is horizontal relative to $\mathbf{v}$ and if we consider a small circle D on S bounding a neighborhood around p, the tangent plane will not intersect D. If p is a saddle point, however, then the tangent plane will intersect such a circle in four distinct points. At a regular point q, the tangent plane divides a small disc neighborhood into two pieces and meets the circle D at two points. In any of these situations, denote by d the number of intersection points. Then we have the following formula for the index:

$$i(p, \mathbf{v}) = 1 - \frac{1}{2}d.$$

This holds for any point on the surface, not just critical points.

Now suppose S is a polyhedral surface in $\mathbb{R}^3$ with V vertices, E edges, and T triangular faces (for example, a triangulated surface; see Appendix B). Note that S need not be convex. We know that $\chi(S) = V - E + T$, and we wish to use Definition 5.1 to deduce a formula for $\chi(S)$ in terms of the indices of critical points for height functions.

Definition 5.2. Let $\mathbf{v}$ be a unit vector in $\mathbb{R}^3$ and let $\pi_{\mathbf{v}}$ be the corresponding projection map. Then $\pi_{\mathbf{v}}$ is *general* for S if $\pi_{\mathbf{v}}(x) \neq \pi_{\mathbf{v}}(y)$ for distinct vertices x and y of S.

Given a general $\pi_{\mathbf{v}}$, a point q on S is *regular* for $\pi_{\mathbf{v}}$ if the plane through q orthogonal to $\mathbf{v}$ cuts the neighborhood $\mathrm{St}(q)$ into two pieces. Note that this implies that any point q in the interior of an edge or face must be regular since no edge or face can be orthogonal for a general $\pi_{\mathbf{v}}$. Vertices, however, can be critical points and we may use the definition above to define an index for them.

Let x be a vertex of S and let D be the boundary of the star $\mathrm{St}(x)$. Note the following fact. If σ is a triangle in $\mathrm{St}(x)$ with vertices x, y, z, then one of the following is true: (i) x lies below both y and z (relative to $\mathbf{v}$); (ii) x lies above both y and z; (iii) x lies between y and z. In the third case, call x a *middle vertex* for σ. Then if we consider the plane through x orthogonal to $\mathbf{v}$, the number of times it intersects D is the number of middle vertices for the triangles in $\mathrm{St}(x)$. We then have the following reformulation of Definition 5.1 for x:

$$i(x, \mathbf{v}) = 1 - \frac{1}{2}(\text{number of } \sigma \text{ with } x \text{ middle for } \sigma).$$

Theorem 5.3. *If $\pi_{\mathbf{v}}$ is general for S, then*

$$\sum_{x \in S} i(x, \mathbf{v}) = \chi(S).$$

Proof. First, note that on a polyhedral surface we have $3T = 2E$. Indeed, since an edge has precisely two triangles in its star, we see that $3T$ is the number of pairs $(\sigma, \text{edge of } \sigma)$, and this is $2E$. With this in mind, we then have

$$\begin{aligned}
\sum_{x \in S} i(x, \mathbf{v}) &= \sum_{x \in S} \left(1 - \frac{1}{2}(\text{number of } \sigma \text{ with } x \text{ middle for } \sigma)\right) \\
&= V - \frac{1}{2} \sum_{x \in S} (\text{number of } \sigma \text{ with } x \text{ middle for } \sigma) \\
&= V - \frac{1}{2} T \text{ (since each } \sigma \text{ has exactly one middle vertex for } \mathbf{v}) \\
&= V - \frac{1}{2}(2E - 2T) \\
&= V - E + T.
\end{aligned}$$

□

Figure 5.1 illustrates Theorem 5.3. The polyhedron S shown has vertices labeled with their indices with respect to the vertical direction $\mathbf{v}$. Note that the sum of the indices is 2, which agrees with $\chi(S)$ (S is homeomorphic to the sphere S^2).

Note that it is possible for a polyhedral surface to have critical points with indices different from ± 1. Consider for example the surface given by the equation $z = x^3 - 3xy^2$. This is often called the "monkey saddle" because the (degenerate) critical point at the origin has three directions of descent (one for each of a monkey's legs and one for its tail). Its graph is shown in Figure 5.2. Note that in a triangulated model, the boundary of the star of the origin intersects the tangent plane (orthogonal to the vector $\mathbf{z}$ spanning the z-axis) in 6 distinct points and hence the index of this point is $i((0,0), \mathbf{z}) = 1 - 6/2 = -2$.

Banchoff also uses these ideas to develop a notion of curvature for embedded polyhedra. It is a fascinating story, but beyond the scope of this book. We refer the interested reader to [Banchoff (1970)] for details.

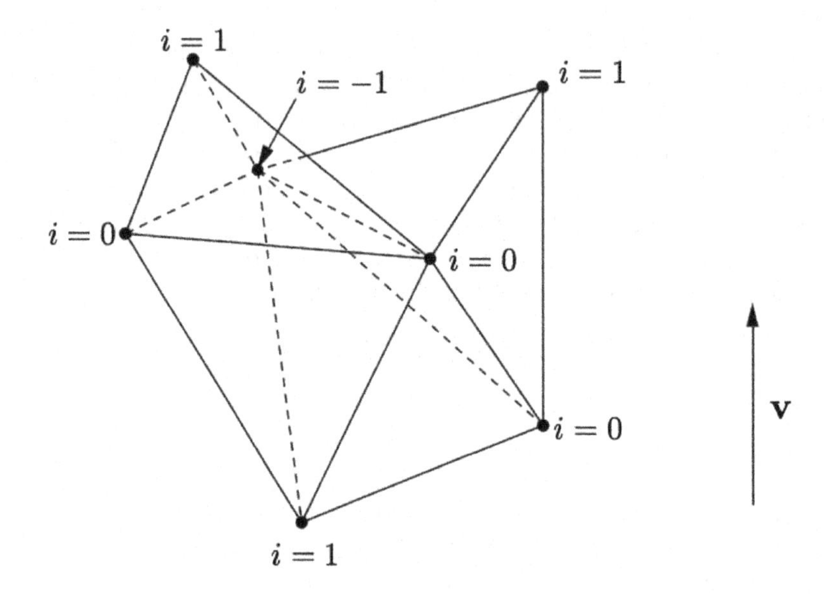

Fig. 5.1 The polyhedron S and the indices of its vertices relative to the vertical direction

5.2 Computational Morse theory

Banchoff's notion of index for vertices of a polyhedral surface embedded in $\mathbb{R}^3$ motivates much of the discrete versions of Morse theory that follow. In this section, we describe a combinatorial version of the Morse–Smale (MS) complex for surfaces (see discussion following Definition 4.11) due to Edelsbrunner, et al. [Edelsbrunner et al. (2003b)].

Recall that the Morse–Smale complex associated to a Morse function f on a manifold M is obtained by taking the intersections of the stable and unstable manifolds of the critical points of f. In general, given two critical points a and b for the function f, the set $W_a^u \cap W_b^s$ is the union of flow lines for f with origin a and destination b; it is a cell of dimension $\text{index}(a) - \text{index}(b)$. Such cells have a particular structure.

Lemma 5.4. *Let M be a closed 2-manifold and let $f : M \to \mathbb{R}$ be a Morse–Smale function. Then each cell in the Morse–Smale complex is a quadrangle whose vertices are critical points of f of indices $0, 1, 2, 1$ in this order. The boundary is possibly glued to itself along vertices and edges.*

Proof. Note that the vertices on the boundary of any region alternate between saddles and other critical points, which also alternate between maxima and minima. The shortest such sequence of vertices is a quadrangle of indices $0, 1, 2, 1$. We claim that a longer sequence would force a critical point in the interior of the

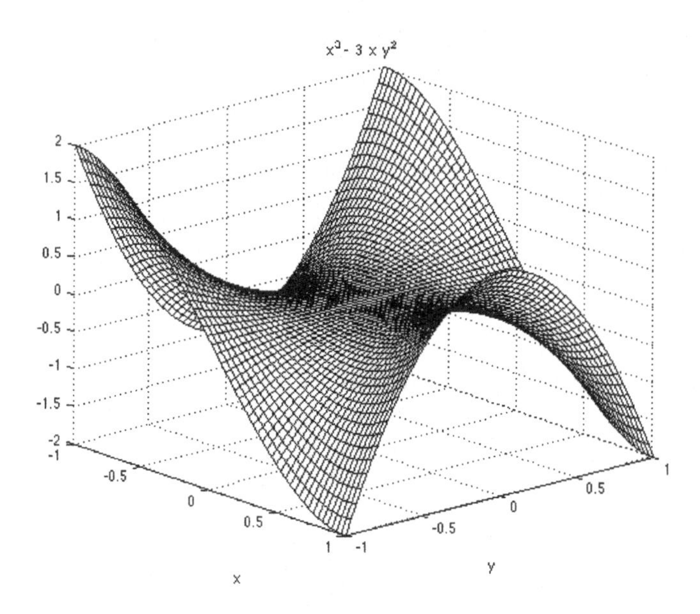

Fig. 5.2 The monkey saddle

region, contrary to the construction of the complex. To this end, suppose we have a boundary cycle of length $4k$, $k \geq 2$. Glue two copies of the region together along their boundaries to form a sphere, gluing each critical point to its copy. Then saddles become regular points while maxima and minima remain as they were. Denote the index of each critical point a by $i(a)$. Then since the Euler characteristic of the sphere is 2, we have

$$\sum_a (-1)^{i(a)} = 2.$$

However, the number of minima and maxima is $2k > 2$, implying there is at least one saddle inside the region. □

There is a combinatorial abstraction of this construction, defined as follows. Let M be a closed surface. A quadrangulation of M is *splittable* if the vertices may be partitioned into three subsets U, V, W, and the arcs may be partitioned into two sets A, B satisfying

(1) $U \cup W$ and V are both independent (a set Z of vertices is *independent* if no two are connected by an arc);
(2) arcs in A have endpoints in $U \cup V$ and arcs in B have endpoints in $V \cup W$;
(3) each vertex $v \in V$ belongs to four arcs, which in cyclic order around v alternate between A and B.

Note that the MS-complex associated to a Morse–Smale function f on M is splittable: U, V, W are the maxima, saddles, and minima, respectively; A connects maxima to saddles and B connects minima to saddles; and saddles have degree four and alternate as required by Lemma 5.4. The MS-complex also satisfies the additional condition that the arcs are paths of steepest ascent of f.

Given a splittable quadrangulation, we may split it into two complexes defined by U, A and W, B. In the case of the MS-complex, these are the complexes of stable and unstable manifolds.

Definition 5.5. A *quasi-MS-complex* of a 2-manifold M and height function f is a splittable quadrangulation whose vertices are the critical points of f and whose arcs are monotonic in f.

This definition is motivated by applications. In practice, we rarely have a smooth function; rather, we have the values of a function on some sampling of the manifold in question. In this situation, the obvious thing to do is to extend the function linearly over the simplices: given $f : M^{(0)} \to \mathbb{R}$, define $f : M \to \mathbb{R}$ by

$$f\left(\sum_i t_i(x)u_i\right) = \sum_i t_i f(u_i)$$

where $t_i(x)$ denotes the barycentric coordinates of the point x in the simplex σ with vertices u_i. Via a small perturbation if necessary, we may assume that f is injective on the set of vertices, and then order the vertices by increasing function values: $f(u_1) < f(u_2) < \cdots < f(u_n)$. Recall that the open star of a vertex is the set $\mathrm{S}(u) = \{\sigma \in M : u \in \sigma\}$ (see Appendix B). Using Banchoff's ideas, define the *lower* and *upper stars* of u by

$$\underline{\mathrm{St}}(u) = \{\sigma \in \mathrm{S}(u) : f(v) \leq f(u), v \in \sigma\}$$
$$\overline{\mathrm{St}}(u) = \{\sigma \in \mathrm{S}(u) : f(v) \geq f(u), v \in \sigma\}.$$

Note that the lower and upper stars partition the complex M:

$$M = \bigcup_u \underline{\mathrm{St}}(u) = \bigcup_u \overline{\mathrm{St}}(u).$$

The lower and upper stars are combinatorial analogues of the stable and unstable manifolds of a smooth Morse function.

The lower and upper stars also allow us to classify vertices by an index. Define a *wedge* as a contiguous section of $\mathrm{S}(u)$ that begins and ends with an edge. Note that $\underline{\mathrm{St}}(u)$ either contains the entire star of u or some number $k+1$ of wedges. The same holds for $\overline{\mathrm{St}}(u)$. If the lower star is the entire star, then $k = -1$ and u is a maximum for f. Similarly, if the upper star is the entire star, then $k = -1$ and u is a minimum for f. Otherwise, if $k = 0$, then u is a regular point; if $k = 1$, then u is a saddle; if $k \geq 2$, then u is called a degenerate, k-fold saddle (such as the monkey saddle).

Given a k-fold saddle u, we can "unfold" it into two saddles of smaller multiplicities, i, j with $i + j = k$. The process is as follows.

Algorithm 5.6. Unfolding multiple saddles

1: **given** a k-fold saddle u, $k \geq 2$
2: split a wedge of $\underline{\mathrm{St}}(u)$ (through a triangle if necessary)
3: split a non-adjacent wedge of $\overline{\mathrm{St}}(u)$ (see Figure 5.3)
4: assign a function value to the new vertex close to the value of $h(u)$

Iterating this procedure, we see that we can replace a k-fold saddle by k regular saddles. This requires $k - 1$ steps, each of which adds an additional vertex to the complex. There is of course some ambiguity in this procedure as we may have many choices of wedge splittings, but we need only choose one in order to proceed.

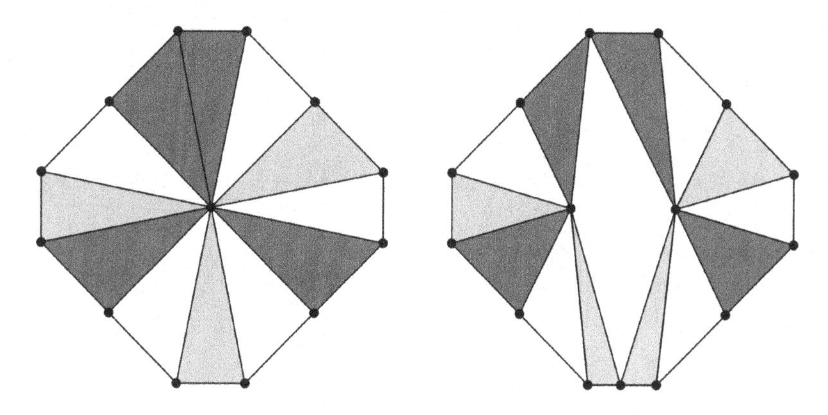

Fig. 5.3 Unfolding a monkey saddle. The darker simplices are in the upper link of the vertex and the lighter shaded simplices are in the lower link.

Recall that the Morse–Smale complex on a manifold is defined using the flow lines of the gradient of the Morse function. These curves do not intersect. In the PL case, however, we do not have a gradient. We do, however, have lines of monotonic decrease of the function f. These curves may merge together or split apart, and they may do so a number of times. To define a quasi-MS-complex, we will simulate differentiability; that is, we will simply pretend that when two such curves merge that they are actually infinitesimally separated. Also, in the smooth case, we require that the stable and unstable manifolds intersect transversely. Non-transversality occurs on a surface when the unstable 1-manifold of a saddle approaches another saddle (e.g., the standard height function on the torus); this may be resolved in the smooth case by taking a small perturbation of the function. In the PL case, non-transversality corresponds to an ascending or descending path ending at a saddle. To resolve this, we will simply extend the path beyond the saddle. We give details below.

Now suppose we have a triangulated compact 2-manifold M with a PL height function $f : M \to \mathbb{R}$. We construct a quasi MS-complex Q in stages. The outline of the algorithm is as follows.

Algorithm 5.7. Quasi MS-complex

1: **given** M, a triangulated 2-manifold, and $f : M \to \mathbb{R}$ a PL height function
2: apply Algorithm 5.8 to produce a complex with junctions
3: apply Algorithm 5.9 to remove junctions and reduce the number of arcs per k-fold saddle
4: apply Algorithm 5.10 to unfold the saddles
5: **return** Q

We now describe the individual components of the algorithm. The first stage builds a complex with vertices corresponding to the critical points, along with some additional vertices corresponding to "junctions." The arcs in the complex connect these vertices in a prescribed manner mimicking the ascending and descending paths of the function.

Algorithm 5.8. Complex with junctions

1: **given** M, a triangulated 2-manifold, and $f : M \to \mathbb{R}$ a PL height function
2: **initialize** $Q = \emptyset$
3: **for all** vertices v in M **do**
4: classify v as a maximum, minimum, or k-fold saddle
5: **if** v is a critical point **then**
6: add v to Q
7: **end if**
8: compute the wedges of $\underline{\mathrm{St}}(v)$ and $\overline{\mathrm{St}}(v)$
9: **for all** wedges in $\underline{\mathrm{St}}(v)$ and $\overline{\mathrm{St}}(v)$ **do**
10: compute the steepest edge in each wedge
11: **end for**
12: **end for**
13: **for all** k-fold saddles v **do**
14: begin $k+1$ ascending and $k+1$ descending paths from v using the steepest edge in each wedge
15: **for all** ascending and descending paths γ beginning at v **do**
16: extend the path using edges of steepest ascent/descent in each wedge
17: **if** the path γ hits a minimum or maximum **then**
18: the path ends
19: add an arc to Q corresponding to γ connecting the critical points at its ends
20: **end if**
21: **if** the path γ hits a previously traced path at a regular point **then**
22: this corresponds to a merging or forking; create a new junction and split the previously traced path, or increase the degree of the previously created junction
23: add a vertex u to Q corresponding to this junction
24: add an arc to Q joining the saddle to u
25: **end if**

26: **if** the path γ hits another saddle **then**
27: add an arc to Q corresponding to γ connecting the saddles
28: **end if**
29: otherwise, the path ends at a regular point; extend it using the edge of steepest ascent/descent in each wedge
30: **end for**
31: **end for**
32: **return** Q

We now remove the junctions, as these are extraneous, and we also eliminate the arcs in Q which end at a saddle point. We do this via the following procedure.

Algorithm 5.9. Extending paths

1: **given** the complex Q
2: order the ascending paths in the order of increasing height; order the descending paths in order of decreasing height
3: **for all** junctions u **do**
4: by definition u corresponds to a regular point of f
5: the first time u is reached along a path, the path continues through u; if it is ascending then one ascending path leaves u through $\overline{\mathrm{St}}(u)$, all ascending paths approach u from $\underline{\mathrm{St}}(u)$, and all descending paths approach u from $\overline{\mathrm{St}}(u)$.
6: duplicate paths ending at u using the orderings
7: concatenate paths in pairs without creating crossings (see Figure 5.4)
8: delete u from Q
9: **end for**
10: **for all** paths γ that end at a saddle v **do**
11: find two cyclically contiguous steepest edges at v
12: find ascending paths approaching v within the wedge of $\underline{\mathrm{St}}(v)$
13: find descening paths approaching v within the wedge of $\overline{\mathrm{St}}(v)$
14: duplicate these paths
15: concatenate the paths in pairs without creating crossings (see Figure 5.5)
16: replace arc corresponding to γ in Q with arcs corresponding to resulting paths
17: **end for**
18: **return** Q

At the end of Algorithm 5.9 the complex Q has only critical points as vertices and its edges correspond to monotonic non-crossing paths from saddles to minima or saddles to maxima. In the case of a k-fold saddle, there are exactly $k+1$ ascending and $k+1$ descending paths beginning at the saddle.

Algorithm 5.10. Eliminating multiple saddles

1: **given** Q from Algorithm 5.9
2: **for all** k-fold saddles, $k \geq 2$ **do**

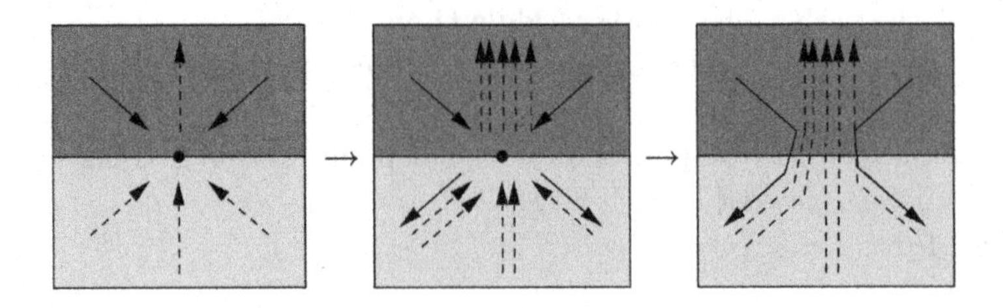

Fig. 5.4 Duplication and concatenation of paths ending at junctions

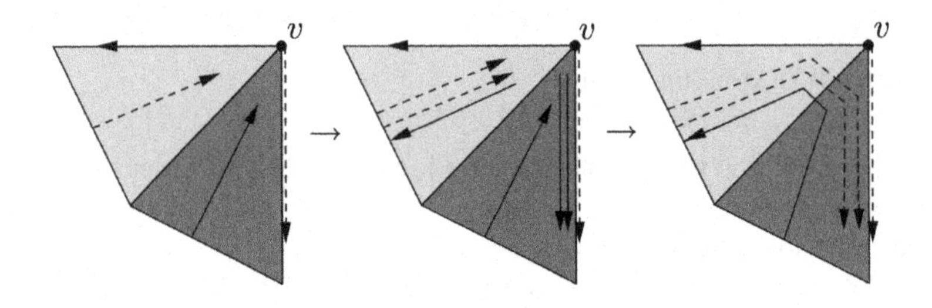

Fig. 5.5 Duplication and concatenation of paths ending at saddles

3: **for all** ascending paths γ **do**
4: duplicate the saddle, the ascending path, and one non-adjacent descending path
5: add the new saddle as a vertex to Q and add an arc to Q corresponding to each path
6: **end for**
7: **end for**
8: **return** Q

The result after Algorithm 5.10 is that we have k vertices for each k-fold saddle and $4k$ paths (4 per new saddle). This process does not create any path crossings in the complex Q.

Proposition 5.11. *Algorithm 5.7 produces a quasi MS-complex for M.*

Proof. The vertices of Q are the unfolded critical points of M; they are therefore maxima, minima, and (simple) saddles. The paths do not cross and Algorithm 5.9 guarantees that the paths go from saddles to maxima or minima. It follows that Q is splittable. Also, the vertices on the boundary of any region in Q alternate between saddles and maxima/minima. Lemma 5.4 then implies that Q is a quadrangulation and hence is a quasi MS-complex. □

The quasi MS-complex Q may be rather large, but it is simpler than the initial triangulation of M. Moreover, the complex Q captures the topology of M in the sense that $H_\bullet(Q) \cong H_\bullet(M)$.

Interlude: persistence canceling

In the smooth case, given a Morse function h, one may cancel critical points in pairs whenever they are connected by a single gradient path. One would like to simplify the quasi MS-complex Q in an analogous manner; the procedure is based on *persistent homology*, which we now describe.

Assume that M is a triangulated 2-manifold without boundary and imagine building the complex M by adding simplices. Order the vertices $u^1, u^2, \ldots, u^n$ by increasing f-value: $f(u^1) < f(u^2) < \cdots < f(u^n)$. For each j, let M^j be the union of the first j lower stars:

$$M^j = \bigcup_{1 \le i \le j} \underline{\mathrm{St}}(u^i).$$

The sequence

$$M^1 \subset M^2 \subset \cdots M^n = M$$

is called the *lower star filtration* of M. For each j, we have three Betti numbers, $\beta_0^j, \beta_1^j, \beta_2^j$. We can compute the Betti numbers of M^{j+1} from those of M^j by examining how the lower star of u^{j+1} connects to M^j. Note that if u^i is a regular vertex, then the Betti numbers of M^i agree with those of M^{i-1}; we therefore need only consider critical vertices. We will use reduced homology and we therefore have $\beta_{-1}^0 = 1$ and $\beta_0^0 = \beta_1^0 = \beta_2^0 = 0$.

Suppose u^{j+1} is a minimum of f. If $j = 0$, then $\beta_{-1}^1 = \beta_{-1}^0 - 1 = 0$. If $j > 0$, then u^{j+1} forms a new component and the 0th Betti number increases by 1: $\beta_0^{j+1} = \beta_0^j + 1$.

If u^{j+1} is a k-fold saddle, then the lower star of u^{j+1} intersects M^j along $k+1$ simple paths. Let $1 \le r \le k+1$ be the number of distinct components touched by these paths. Then the addition of $\underline{\mathrm{St}}(u^{j+1})$ decreases the number of components: $\beta_0^{j+1} = \beta_0^j - (r-1)$ and increases the number of non-bounding cycles: $\beta_1^{j+1} = \beta_1^j + (k+1-r)$.

Finally, if u^{j+1} is a maximum, then if $j+1 = n$ the addition of $\underline{\mathrm{St}}(u^{j+1})$ completes the manifold and $\beta_2^n = \beta_2^{n-1} + 1 = 1$. Otherwise, the lower star fills in a hole and we decrease the first Betti number: $\beta_1^{j+1} = \beta_1^j - 1$.

In each case, we see that critical vertices either create new homology classes (minima, saddles, and maxima) or destroy existing homology classes (saddles and maxima). Call a critical vertex *positive* if its addition creates a cycle and *negative* if its addition destroys a cycle. Note that these acts often occur in pairs; that is, a homology class created by the addition of a positive critical point may be destroyed later in the filtration by a negative critical point. This is the idea behind *persistence.*

Definition 5.12. The *persistence pairing* of critical vertices is defined as follows. Proceed through the lower star filtration in increasing order. A positive minimum creates a new component represented by the vertex. A negative saddle connects two components of the existing complex, each of which is represented by its lowest minimum; the saddle vertex is paired with the higher of the minima and the other (lower) minimum remains unpaired, representing the merged component. A positive saddle creates and represents a new non-bounding 1-cycle. A negative maximum fills in such a 1-cycle, which is homologous to a sum of cycles, each represented by a positive saddle; the maximum is paired with the highest of these saddles, and the other saddles remain unpaired as representatives of their cycles. At the end of this process, we have pairs of the type (minimum, saddle) and (saddle, maximum) and a collection of $\beta_0 = 1$ minima, β_1 simple saddles, and $\beta_2 = 1$ maxima that remain unpaired. The *persistence* of a critical point x is $p(x) = |f(x) - f(y)|$, if x is paired with y, and $p(x) = \infty$ if x is unpaired.

Remark 5.13. Persistence is an important concept in computational topology, and the reader is encouraged to learn more. While it is beyond the scope of this book, a quick reading of the basics of the theory of persistent homology should make it clear that there is an intimate relationship between it and the discrete Morse theory of Part II. We refer the reader to [Edelsbrunner and Harer (2009)] for an elementary introduction to persistence.

With persistence in hand, we now describe a simplification procedure for the quasi MS-complex. An example to keep in mind is that of a cubic function $q : \mathbb{R} \to \mathbb{R}$. Generically, such a map q will have one local maximum and one local minimum; the corresponding critical points are paired. Now imagine deforming the map q by sliding the maximum toward the minimum. When the points meet they form a single degenerate critical point (think of the origin for $q(x) = x^3$) before disappearing completely. This requires a local deformation of q near the points. In general, we want a procedure to cancel positive minima with negative saddles and positive saddles with negative maxima.

Suppose then that b is a minimum vertex paired with the saddle a and these are such that ab is an arc in the quasi MS-complex Q. Then a is connected to another minimum c and to two maxima d and e. We simplify M as follows: delete the two ascending paths from a to d and e, and contract the two descending paths from a to b and c. This pulls a and b into c, which inherits the connections of b. This procedure is called the *cancellation* of a and b. There is also the symmetric case where b is a maximum.

Note that it might be the case that $d = e$ and $b = c$. These cannot occur at the same time, however, and we prohibit the cancellation in the $b = c$ case since it would change the topology of the underlying manifold.

We perform cancellations in the order of increasing persistence. It is possible, *a priori*, that critical pairs are not adjacent in the quasi MS-complex, but that is handled by the following result.

Lemma 5.14. *For every $i > 0$, the ith pair of critical points ordered by persistence* *forms an arc in the complex obtained by canceling the first $i-1$ pairs.*

Proof. We may assume without loss of generality that the ith pair consists of a negative saddle $a = u^{j+1}$ and a positive minimum b. The vertex b is in some component of M^j and one of the descending paths originating at a enters this component. Because the path cannot ascend it eventually ends at some minimum z in the same component. Either $z = b$, in which case we are done, or z has already been paired with a saddle $e \neq a$. In this case, e has height less than a, it belongs to the same component of M^j as b and z, and z, e is one of the first $i-1$ pairs of critical points. This implies that when z gets canceled, the path from a to z is extended to another minimum d, which again belongs to the same component. In the end, all minima in the component other than b get canceled, which implies that the initial path from a to z gets extended all the way to b. This proves the result. □

So, given a quasi MS-complex, we may apply the persistence canceling procedure to simplify it further by canceling pairs of critical simplices. This may speed up homology calculations since we have a smaller complex to consider.

One might hope to extend the ideas of this section to higher-dimensional spaces. While it is possible to say something in the 3-dimensional case [Edelsbrunner et al. (2003a)], the explosion of the number of possible configurations of critical points in higher dimensions makes the problem intractable.

5.3 Another version of PL-Morse theory

Finally, we present a version of Morse theory on simplicial complexes developed by Bestvina and Brady (see [Bestvina and Brady (1997)],[Bestvina (2008)]). Let X be a (finite) simplicial complex with $\dim X = m$ and for each p-simplex σ, denote by $\chi_\sigma : \Delta^p \to X$ the characteristic function of σ (i.e., χ_σ is an embedding of the standard p-simplex Δ^p (which is itself embedded in $\mathbb{R}^m$), and its restriction to any face of Δ^p agrees with the characteristic function of another simplex precomposed with an affine homeomorphism of $\mathbb{R}^m$). The theory we now describe is true for more general complexes (cubical complexes, for example), but we restrict our attention here to the simplicial case.

Definition 5.15. Let X be a finite simplicial complex. A function $f : X \to \mathbb{R}$ is a *Morse function* if the following conditions hold.

(1) For each simplex σ the composition $f \circ \chi_\sigma : \Delta^p \to \mathbb{R}$ is the restriction of an affine function $\mathbb{R}^m \to \mathbb{R}$.
(2) If $f|_\sigma$ is constant, then $\dim \sigma = 0$ (that is, there are no "horizontal" cells).
(3) The image of $X^{(0)}$, the 0-skeleton of X, under f is a discrete subset of $\mathbb{R}$.

Remark 5.16. We stated condition (1) above in that form so that it is clear how one would generalize this definition to more general complexes. For a simplicial complex, however, this condition is equivalent to insisting that the restriction of f to each cell is linear in the barycentric coordinates. That is, if $\sum t_i v_i$ is a point in a simplex σ with vertices $\{v_i\}$, then

$$f\Big(\sum t_i v_i\Big) = \sum t_i f(v_i).$$

An immediate consequence of the definition is that the only critical points of such a Morse function are vertices of the complex of X.

Proposition 5.17. *Suppose $J = [a,b]$ is a closed interval in $\mathbb{R}$ such that $(a,b] \cap f(X^{(0)}) = \emptyset$. Then $f^{-1}(J)$ deformation retracts to $f^{-1}(a)$.*

Proof. For each i, set $A_i = f^{-1}(a) \cup (f^{-1}(J) \cap X^{(i)})$. Then the A_i give an increasing filtration of $f^{-1}(J)$, with $A_{-1} = f^{-1}(a)$. It therefore suffices to show that for each i, A_{i+1} deformation retracts to A_i. Suppose σ is an $(i+1)$-simplex and use χ_σ to identify it with $\Delta^{i+1} \subset \mathbb{R}^m$. We may assume that f is a height function on σ from which it follows that $\sigma \cap f^{-1}(J)$ is a convex polytope. There are two horizontal faces: the bottom face which maps to a (which may be a vertex if σ has a vertex mapping to a) and the top face which maps to b. Consider the radial deformation from a point just above the top face; this gives a deformation retraction from $\sigma \cap f^{-1}(J)$ to its boundary minus the top face. Assembling these deformation retractions over all $(i+1)$-simplices gives the required deformation retraction from A_{i+1} to A_i. □

Proposition 5.17 implies that the homotopy type of X changes only at vertices. We have the following analogues of the upper and lower discs discussed in Section 3.3.

Definition 5.18. Let v be a vertex of X. A simplex σ containing v is *descending* if $f|_\sigma$ attains its maximum at v. The *descending link* at v, $\mathrm{Lk}_\downarrow(v,X)$ is the link of v in the union of all descending simplices. Replacing maximum by minimum, we have the corresponding definition of *ascending simplices* and the *ascending link* $\mathrm{Lk}_\uparrow(v,X)$.

Proposition 5.19. *Let $J = [a,b]$ be a closed interval in $\mathbb{R}$ and suppose $f^{-1}(J)$ contains one vertex v. Assume $f(v) = b$. Then the pair $(f^{-1}(J), f^{-1}(a))$ is homotopy equivalent to $(Q, f^{-1}(a))$ relative to $f^{-1}(a)$, where Q is $f^{-1}(a)$ with the cone on $\mathrm{Lk}_\downarrow(v,X)$ attached.*

Proof. Let σ be a descending simplex at v and let $S_\sigma = \sigma \cap f^{-1}([a,b])$. Then S_σ is the cone on $\mathrm{Lk}_\downarrow(v,\sigma) = \mathrm{Lk}(v,\sigma)$ with cone point v and base contained in $f^{-1}(a)$. Let A_{-1} be the union of $f^{-1}(a)$ and all the S_σ over the descending simplices σ. Then A_{-1} is $f^{-1}(a)$ with the cone on $\mathrm{Lk}_\downarrow(v,X)$ attached. Setting $A_i = A_{-1} \cup (f^{-1}(J) \cap X^{(i)})$, we see that A_{i+1} deformation retracts to A_i, as in the proof of Proposition 5.17. □

Remark 5.20. Proposition 5.19 holds if there is more than one vertex in $f^{-1}(b)$: the cones are attached in a pairwise disjoint manner. Also, if the vertices of $f^{-1}(J)$ lie in $f^{-1}(a)$, then $f^{-1}(J)$ deformation retracts to $f^{-1}(b)$ with cones on the ascending links of vertices in $f^{-1}(a)$ attached.

As an application of these ideas we present the following example. For more interesting examples focused on applications to finiteness properties of groups, we refer the reader to [Bestvina (2008)],[Bestvina and Brady (1997)].

Example 5.21. Let K be a field and denote by $SL_n(K)$ the group of $n \times n$ matrices over K with determinant 1. Define a simplicial complex X as follows. The vertices of X are the subspaces of K of dimension $d = 1, 2, \ldots, n-1$. An i-simplex in X exists for every flag

$$V_1 \subset V_2 \subset \cdots \subset V_{i+1}$$

of subspaces. A maximal simplex therefore has dimension $n-2$. The complex X is called the (spherical) *Tits building* for $SL_n(K)$.

Theorem 5.22 (Solomon-Tits). *X has the homotopy type of a wedge of $(n-2)$-spheres.*

Proof. We proceed by induction on n. The case $n = 2$ follows immediately as $X = \mathbb{P}^1_K$ and this is a discrete set.

Suppose $n > 2$ and fix a line ℓ in K^n. Define a map

$$f : X^{(0)} \to \{0,1\} \times \{1, 2, \ldots, n-1\}$$

by $f(V) = (a, \dim V)$, where $a = 0$ if $\ell \subset V$ and $a = 1$ otherwise. Order $\{0,1\} \times \{1, \ldots, n-1\}$ lexicographically. Note that we could extend f to a map into $\mathbb{R}$ by embedding its image into $\mathbb{R}$ preserving the order, but this is unnecessary. The absolute minimum of f is $(0,1)$, realized only by ℓ. Note that adjacent vertices map to distinct points.

We now describe the descending links. Let V be a vertex and say $f(V) = (a, \dim V)$. There are three cases.

(1) If $a = 0$ and $V \neq \ell$, then $\mathrm{Lk}_{\downarrow}(V, X)$ is the complex of proper subspaces of V containing ℓ. This is a cone with cone point ℓ.
(2) If $a = 1$ and $\dim V < n-1$, then $\mathrm{Lk}_{\downarrow}(V, X)$ has two types of vertices: proper subspaces of V and proper subspaces of K^n that contain W, the span of $V \cup \ell$. This is a cone with cone point W.
(3) If $a = 1$ and $\dim V = n-1$, then $\mathrm{Lk}_{\downarrow}(V, X)$ consists of proper subspaces of V, which is the Tits building for $SL_{n-1}(K)$.

The result now follows from Proposition 5.19 and the inductive assumption. □

5.4 Exercises

(1) Consider each of the five Platonic solids (tetrahedron, cube, octahedron, dodecahedron, and icosahedron) and embed them in $\mathbb{R}^3$ so that each vertex is at a different height relative to the z-axis. Compute the index of each vertex and verify Theorem 5.3.
(2) Consider again the five Platonic solids, embedded in $\mathbb{R}^3$ as in the previous exercise. With these vertex orderings, construct the associated quasi MS-complexes. How many saddles do you obtain? Perform persistence canceling to simplify the complexes as much as possible.
(3) Geographic data is available online (for example, from the NOAA at `http://www.ngdc.noaa.gov/mgg/topo/globe.html`). If you have access to an implementation of the algorithms of Section 5.2 (or if you can write one yourself), download some of these data sets and analyze the associated height functions.
(4) The Solomon–Tits Theorem asserts that the spherical Tits building for $SL_n(K)$ is a wedge of $(n-2)$-spheres. If K is a finite field, compute the number of spheres in the wedge.

Bibliographic notes

There are other versions of PL Morse Theory aside from those presented in this chapter. In particular, we refer the interested reader to the papers of Kosiński [Kosinski (1962)] and Kearton and Lickorish [Kearton and Lickorish (1972)].

PART 2
Discrete Morse Theory

Chapter 6

First Steps

We now leave the smooth realm behind and take a much different approach. The category of smooth manifolds and smooth maps enjoys many nice properties, and we have seen that Morse theory is a powerful tool for the analysis of the topology of manifolds. However, in many areas of mathematics and in physical applications, we often do not have the luxury of considering smooth manifolds. In combinatorics, for example, there are many situations in which simplicial complexes arise naturally. Data analysis is another such setting. One would like to understand the topology of these complexes and a Morse-type theory would be very useful for such an analysis. We met some piecewise linear methods in Chapter 5, but these have some drawbacks.

In this part of the book, we introduce the discrete Morse theory developed by R. Forman. This theory works in great generality on cell complexes and we will be able to deduce analogues of the main theorems of smooth Morse theory in this context. Some of these become almost trivial in the discrete setting (e.g., handle cancellation), but we shall see that behind this simplicity lies a great deal of hidden computational complexity.

6.1 Basic definitions

In most applications of discrete Morse theory, it is sufficient to restrict attention to simplicial complexes. However, for completeness, we present the basic definitions in the greatest possible generality. Recall that a CW-complex X is *regular* if for each cell $\sigma^{(p)}$ the attaching map $f_\sigma : \partial\sigma^{(p)} \to X^{(p-1)}$ is a homeomorphism onto its image.

Example 6.1. The canonical example of a non-regular complex is the n-sphere S^n with the CW-structure consisting of a single vertex and a single n-cell. A regular structure on S^n may be obtained inductively by starting with S^0, the space of two points, and then viewing S^{i+1} as the equator S^i with two $(i+1)$-cells attached as the northern and southern hemispheres. The resulting CW-structure has two cells in each dimension $0, 1, \ldots, n$.

One advantage of using regular CW-complexes is that the calculation of incidence numbers of cells is easy. If $\sigma^{(p)} < \tau^{(p+1)}$, and we fix an orientation on the cells of X, then

$$\langle \partial\tau, \sigma \rangle = \pm 1,$$

depending on whether the orientation of σ agrees with the induced orientation from τ or not ([Massey (1991)], p. 244). Simplicial complexes enjoy this property, for example. In particular, when using a regular CW-decomposition of X, the boundary matrices in the chain complex $C_\bullet(X)$ have entries $0, \pm 1$. A disadvantage of regular complexes is that such a decomposition typically requires more cells than is strictly necessary (the sphere example above shows this rather dramatically).

In a general CW-complex, some regularity may exist. The following definition makes this precise.

Definition 6.2. Suppose $\sigma^{(p)} < \tau^{(p+1)}$ and let $h : e^{(p+1)} \to X$ be the continuous map taking the interior of $e^{(p+1)}$ homeomorphically to τ. Then σ is a *regular face* of τ if

(1) $h : h^{-1}(\sigma) \to \sigma$ is a homeomorphism;
(2) $\overline{h^{-1}(\sigma)}$ is a closed p-ball.

Thus a regular CW-complex is a complex in which all faces are regular.

We are now ready to define discrete Morse functions.

Definition 6.3. Let X be a finite CW-complex and denote by K_p the set of p-cells of X. We denote by K the union of the K_p. A *discrete Morse function* on X is a function

$$f : K \to \mathbb{R}$$

satisfying the following for all $\sigma \in K_p$:

(1) If σ is an irregular face of $\tau^{(p+1)}$, then $f(\tau) > f(\sigma)$. Also,
$$\#\{\tau^{(p+1)} > \sigma : f(\tau) \leq f(\sigma)\} \leq 1.$$
(2) If $v^{(p-1)}$ is an irregular face of σ, then $f(v) < f(\sigma)$. Also,
$$\#\{v^{(p-1)} < \sigma : f(v) \geq f(\sigma)\} \leq 1.$$

Heuristically, a discrete Morse function is a function on the cells of X that increases generically with the dimensions of the cells. That is, for any given cell there is at most one coface with a smaller value and at most one face with a larger value.

Remark 6.4. For notational convenience, we will often simply write discrete Morse functions on X as $f : X \to \mathbb{R}$ with no mention of the set K of cells of X. This notational convention will not cause confusion.

Definition 6.5. Let f be a discrete Morse function on X. A cell $\sigma^{(p)}$ is a *critical cell of index* p if the following two conditions hold:

(1)
$$\#\{\tau^{(p+1)} > \sigma : f(\tau) \leq f(\sigma)\} = 0;$$

(2)
$$\#\{v^{(p-1)} < \sigma : f(v) \geq f(\sigma)\} = 0.$$

Let us pause to consider why these definitions make good sense from a topological point of view. Suppose X is a triangulated oriented compact surface without boundary and that f is a discrete Morse function on it. Any edge is a face of exactly two triangles. Suppose e is a critical edge. Then the function value at the endpoints of e must be smaller than $f(e)$; that is, if one tries to exit e through an end then one must go "downhill." Similarly, the function value on the triangles adjacent to e must be larger than $f(e)$ so that one travels "uphill" in those directions. An index 1 critical point for a Morse function on the surface has similar properties: one moves downhill along an integral curve passing through the critical point and uphill along an orthogonal curve (in some local coordinate system). Thus, it is useful to imagine that if σ is a critical cell of index p for some discrete Morse function on a triangulated manifold M, then there is a smooth Morse function on M with a critical point of index p in the interior of σ. One can prove a rigorous result to this effect; see e.g. [Gallais (2010)].

Example 6.6. Let X be any CW-complex. Then the function f defined by

$$f(\sigma) = \dim \sigma$$

is a discrete Morse function on X. Note that *every* cell is critical. This is the discrete analogue of the constant function on a manifold, for which every point is critical (constant functions are not Morse functions, of course). Thus we see that discrete Morse functions exist on any CW-complex.

Example 6.7. A less trivial example is obtained as follows. Denote by K_0 the set of vertices of X and assign distinct positive values to elements of K_0 arbitrarily (i.e., choose an injection $f : K_0 \to \mathbb{R}^+$). Then for any $p > 0$ define

$$f(\sigma^{(p)}) = \sum_{\alpha^{(p-1)}<\sigma} f(\alpha).$$

We claim this is a discrete Morse function on X. Indeed, if $\tau^{(p+1)} > \sigma^{(p)}$, then

$$\begin{aligned} f(\tau) &= \sum_{\alpha^{(p)}<\tau} f(\alpha) \\ &= f(\sigma) + \sum_{\alpha\neq\sigma} f(\alpha) \\ &> f(\sigma); \end{aligned}$$

and if $\alpha^{(p-1)} < \sigma$, then

$$\begin{aligned} f(\sigma) &= \sum_{\beta^{(p-1)}<\sigma} f(\beta) \\ &= f(\alpha) + \sum_{\beta \neq \alpha} f(\beta) \\ &> f(\alpha). \end{aligned}$$

Example 6.8. The function in Figure 6.1 is a discrete Morse function on the boundary of the 2-simplex. The critical cells are $f^{-1}(0)$ and $f^{-1}(5)$.

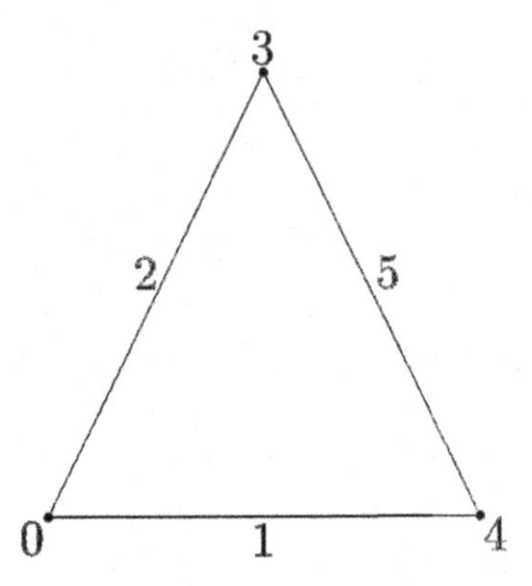

Fig. 6.1 A discrete Morse function on the boundary of the 2-simplex

Example 6.9. A triangulation of the torus is shown in Figure 6.2 (identify the top and bottom edges, and the two side edges). The reader should verify that the numerical assignment shown on the simplices is a discrete Morse function on the torus.

Example 6.10. A critical point of index 0 for a discrete Morse function f (i.e., a critical vertex) must be a local minimum of f. Similarly, a critical cell of index n ($n = \dim X$) is a local maximum of f.

Lemma 6.11. *Suppose σ is a p-cell that is not critical for a discrete Morse function f. Then exactly one of the following conditions holds:*

(1) There is a $(p+1)$-cell τ with $f(\tau) \leq f(\sigma)$;
(2) There is a $(p-1)$-cell α with $f(\alpha) \geq f(\sigma)$.

Proof. Since σ is not critical at least one of the conditions holds. Suppose both conditions hold. Then we must have $p \geq 1$ by the second condition. Moreover, the first condition implies that σ is a regular face of τ. If $\nu \neq \sigma$ is any other p-face of τ, then $f(\nu) < f(\tau)$, and hence $f(\nu) < f(\sigma)$. Also, α is a regular face of σ. We claim there is a p-cell $\mu \neq \sigma$ with $\tau > \mu > \alpha$. Given this, since $f(\alpha)$ cannot be greater

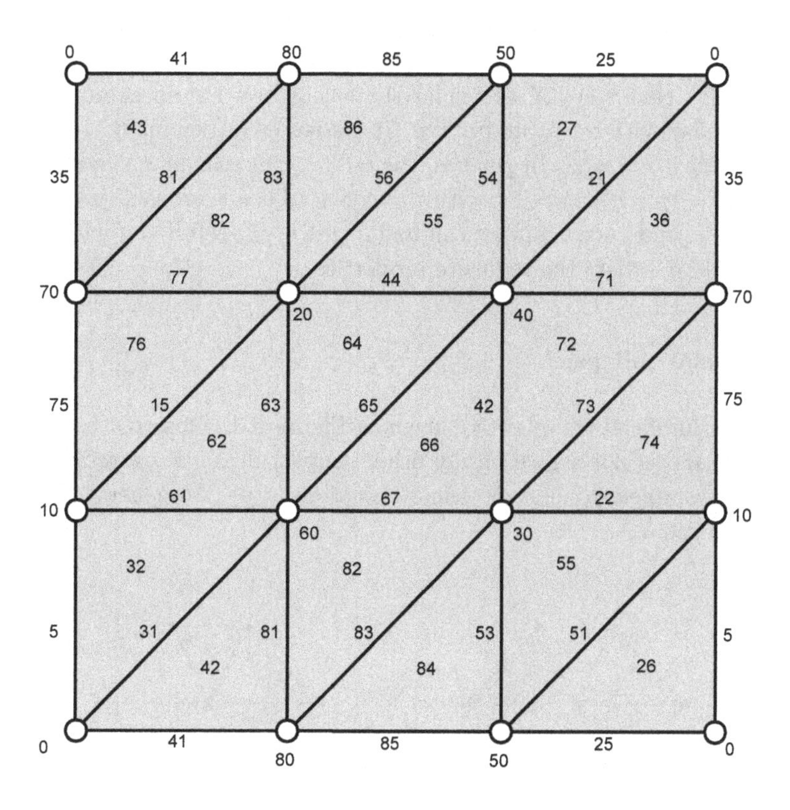

Fig. 6.2 A discrete Morse function on the torus

than or equal to both $f(\sigma)$ and $f(\mu)$ we must have $f(\alpha) < f(\mu)$. We then have the following chain of inequalities:

$$f(\sigma) \leq f(\alpha) \leq f(\mu) < f(\tau) \leq f(\sigma).$$

This is a contradiction.

The existence of the cell μ is left as an exercise. □

We shall need the following technical result in the next chapter.

Lemma 6.12. *Let σ be a p-cell in a regular cell complex X and suppose $\tau > \sigma$. Then there is a $(p+1)$-cell $\tilde{\tau}$ with $\sigma < \tilde{\tau} \leq \tau$ with $f(\tilde{\tau}) \leq f(\tau)$.*

Proof. We have $\dim \tau > \dim \sigma$. If $\dim \tau = p + 1$ we can take $\tilde{\tau} = \tau$. Suppose $\dim \tau = p + r$ for some $r > 1$. We claim that there are two $(p + r - 1)$-faces ν_1 and ν_2 satisfying $\tau > \nu_1 > \sigma$ and $\tau > \nu_2 > \sigma$. Assuming this for the moment, we must then have either

$$f(\nu_1) < f(\tau)$$

or

$$f(\nu_2) < f(\tau).$$

The statement of the lemma then follows by induction.

It remains to verify the claim. We proceed by induction on r. If $r = 1$, that is, $\tau^{(p+1)} > \nu^{(p-1)}$, then since X is regular the p-cells in τ are dense in $\overline{\tau} - \tau$. Thus, there is a p-cell σ with $\tau > \sigma$ and $\sigma > \nu$. It follows as in Lemma 6.11 that there is a $\tilde{\sigma}^{(p)} \neq \sigma$ with $\tau > \tilde{\sigma} > \nu$. In general, the $(p+r-1)$-cells in $\overline{\tau} - \tau$ are dense and so we can find a $(p+r-1)$-cell σ with $\tau > \sigma > \nu$. Then there is a $(p+r-2)$-cell $\tilde{\nu}$ with $\sigma > \tilde{\nu} > \nu$, and once again we can find a $(p+r-1)$-cell $\tilde{\sigma} \neq \sigma$ with $\tau > \tilde{\sigma} > \tilde{\nu}$. The cells σ and $\tilde{\sigma}$ satisfy the requisite properties. □

6.2 Simplicial collapses

Consider the simplicial complex X shown in Figure 6.3. Observe that the edge σ is a face of τ and is not a face of any other simplex in X. If we remove σ and τ, the resulting complex X' has the same homotopy type. In general, we have the following definition.

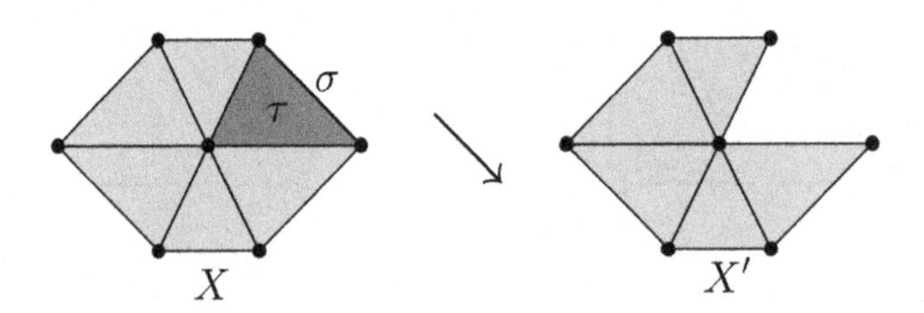

Fig. 6.3 The complexes X and X'

Definition 6.13. Let X be a CW-complex and suppose $\sigma^{(p)} < \tau^{(p+1)}$ are cells such that (*i*) σ is a regular face of τ, and (*ii*) σ is not a face of any other cell of X. Let $X' = X - (\sigma \cup \tau)$. Then we say that X *collapses onto* X'. If X can be transformed into Y by a sequence of these operations, we write $X \searrow Y$. In this case, Y is a deformation retract of X.

The equivalence relation generated by collapses is called *simple homotopy equivalence.* This relation is strictly finer than ordinary homotopy equivalence; that is, there exist simplicial complexes which are homotopy equivalent but not simple homotopy equivalent.

As a preview of results to come, recall the case of a smooth Morse function $f : M \to \mathbb{R}$ and an interval $[a, b]$ such that $f^{-1}([a, b])$ has no critical points of f. In Theorem 3.1 we showed that the sublevel set M_a is a deformation retract of M_b. The proof was to consider the flow lines associated to a gradient-like vector field for f and let the boundary of M_a flow along these lines to the boundary of

M_b. The deformation retraction is given by reversing this process. Now imagine the sublevel set M_b as a triangulated manifold with boundary such that M_a is a triangulated submanifold. The free faces in ∂M_b are codimension-1 faces of n-cells in M_b and they are not faces of any other n-simplices in M_b. We may therefore collapse these, one at a time, to obtain a new complex $M_b' \subset M_b$, whose boundary is homeomorphic to that of M_b. We then collapse the free faces and their adjacent n-simplices. Iterating this process, we eventually get to M_a so that $M_b \searrow M_a$. See Figure 6.4 for an illustration.

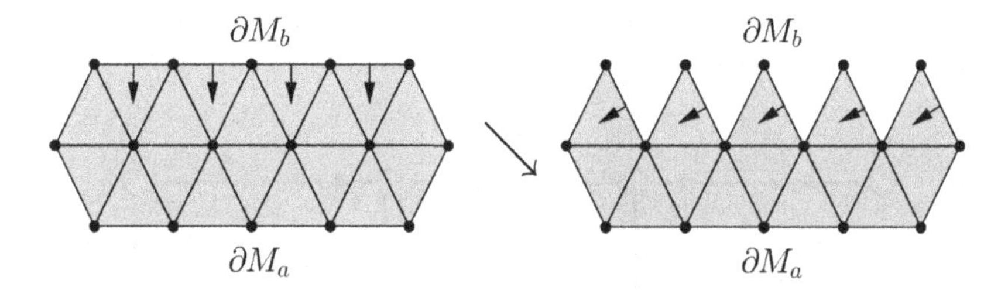

Fig. 6.4 The collapse $M_b \searrow M_a$. The arrows in the left figure show the collapses involving free faces on ∂M_b. The arrows in the right figure show the next collection of collapses. The next collapses would remove the diagonal edges, and so on, until ∂M_a is all that remains.

The question is then whether or not we can perform the analogous operations with a discrete Morse function on a CW-complex. This will be the subject of Chapter 7.

Describing a discrete Morse function requires verification of several local combinatorial conditions. Writing down such a function on an arbitrary simplicial complex which has relatively few critical simplices is tricky. Indeed, consider the discrete Morse function on the torus from Figure 6.2. Now imagine beginning with no values assigned to any simplex and trying to attach numbers in such a way that the defining conditions for a discrete Morse function are satisfied in a nontrivial manner (i.e., by not simply making each simplex critical).

In the smooth case, recall that many of the fundamental theorems relating the topology of the manifold to the critical points of a Morse function were proved using a gradient-like vector field rather than the function itself. In fact, the values of the Morse function at the critical points are generally irrelevant and can be moved at will without altering the diffeomorphism type of the manifold. Functions, then, are *quantitative* in nature and tell us about the geometry of the space, while vector fields are *qualitative* and provide information about the topology of the space. They both have their uses, but more often than not we employ the vector field associated to a Morse function rather than the function itself.

To that end, in the next section we define the gradient vector field associated to a discrete Morse function and prove results recasting discrete Morse theory as questions in the theory of directed graphs.

6.3 Discrete vector fields

We begin with the general notion of a discrete vector field and then characterize those which are gradients. Throughout this chapter, X will denote a regular CW-complex. The definitions given below may be generalized to an arbitrary CW-complex, but regularity will suffice for our purposes.

Definition 6.14. A *discrete vector field* V on X is a collection of pairs $\{\alpha^{(p)} < \beta^{(p+1)}\}$ of cells in X such that no cell is in more than one pair.

We visualize a discrete vector field by drawing an arrow from $\alpha^{(p)}$ to $\beta^{(p+1)}$. An example of a discrete vector field on the torus is shown in Figure 6.5.

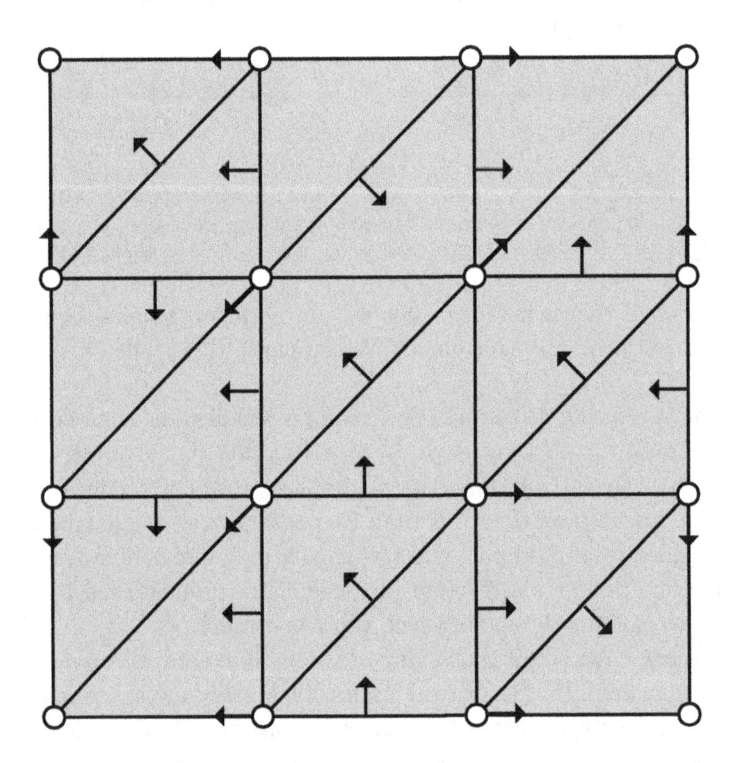

Fig. 6.5 A discrete vector field on the torus

The condition that no cell belongs to more than one pair in V implies that exactly one of the following is true for each cell α:

(1) α is the head of exactly one arrow;
(2) α is the tail of exactly one arrow;
(3) α is neither the head nor tail of an arrow.

The last condition occurs when α is not paired with another cell; in this case we call α a *critical cell* of V. We note the following result.

Theorem 6.15. *Let V be a discrete vector field on X. Then*

$$\chi(X) = \sum_{p=0}^{\dim X} (-1)^p \#\{\text{critical cells of } V \text{ of dimension } p\}.$$

Proof. We know that $\chi(X)$ is the alternating sum of the number of cells in X. However, given V, a cell which is not critical is paired with another cell of dimension one lower or higher. These cells cancel pairwise in the sum

$$\sum_{p=0}^{\dim X} (-1)^p \#\{p\text{-cells of } X\},$$

leaving only the critical cells. □

Discrete vector fields clearly exist. The simplest example is the empty field $V = \emptyset$. In this vector field all cells are critical. A simple example is shown in Figure 6.6. Another example is provided by the following definition.

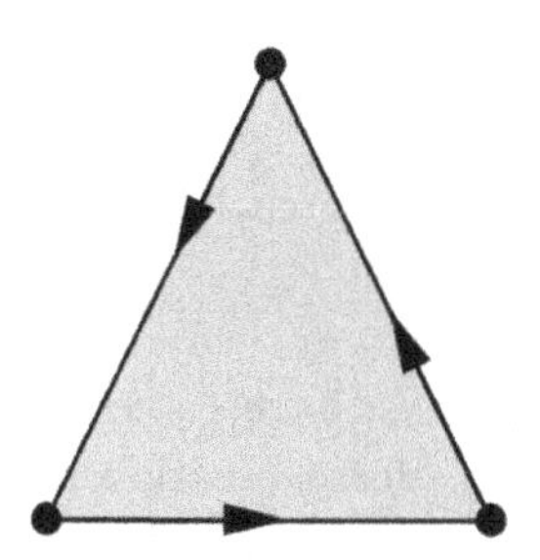

Fig. 6.6 A discrete vector field on the 2-simplex

Definition 6.16. Let X be a regular CW-complex and suppose f is a discrete Morse function on X. The *gradient vector field of* f, $-\nabla f$, is defined as follows. If σ is a critical cell for f, then σ is critical for $-\nabla f$. If $\alpha^{(p)}$ is not a critical cell for f, then by Lemma 6.11 there is either a $(p+1)$-cell $\tau > \alpha$ with $f(\tau) \leq f(\alpha)$ or a $(p-1)$-cell $\mu < \alpha$ with $f(\mu) \geq f(\alpha)$ (and exactly one of these conditions holds). In the former case, $\{\alpha < \tau\} \in -\nabla f$; in the latter, $\{\mu < \alpha\} \in -\nabla f$.

The vector field on the torus shown in Figure 6.5 is the gradient of the discrete Morse function of Figure 6.2.

We use the notation $-\nabla f$ for the gradient because the vectors drawn on X point in the direction of *decrease* of f. Note that for any CW-complex X, the discrete Morse function $f(\sigma) = \dim \sigma$ has $-\nabla f = \emptyset$.

We now discuss the analogue of integral curves for discrete vector fields.

Definition 6.17. Let V be a discrete vector field on X. A *V-path* is a sequence of cells

$$\alpha_0^{(p)} < \beta_0^{(p+1)} > \alpha_1^{(p)} < \beta_1^{(p+1)} > \cdots < \beta_{r-1}^{(p+1)} > \alpha_r^{(p)},$$

where each pair $\alpha_i^{(p)} < \beta_i^{(p+1)}$ is in V and $\alpha_i^{(p)} \neq \alpha_{i+1}^{(p)}$ for $i = 0, \ldots, r-1$. The path is *closed* if $r \geq 1$ and $\alpha_0 = \alpha_r$. The *length* of the path is r. The path is *non-trivial* if $r > 0$.

Recall that in the case of smooth vector fields on manifolds, gradient-like fields have no recurrent dynamics. That is, the integral curves have no loops; once a curve leaves a point it does not return. A similar observation characterizes those discrete vector fields which are the gradient of a discrete Morse function.

Lemma 6.18. *Suppose that $V = -\nabla f$ for a discrete Morse function f on X. Then a sequence of simplices $\alpha_0 < \beta_0 > \alpha_1 < \beta_1 > \cdots < \beta_{r-1} > \alpha_r$ is a V-path if and only if*

$$f(\alpha_0) \geq f(\beta_0) > f(\alpha_1) \geq f(\beta_1) > \cdots \geq f(\beta_{r-1}) > f(\alpha_r).$$

Proof. Suppose the given sequence is a V-path. Then by definition we have $f(\alpha_i) \geq f(\beta_i)$ for $i = 0, \ldots, r$. Moreover, since each α_{i+1} is not paired with β_i, we must have $f(\beta_i) > f(\alpha_{i+1})$.

Conversely, suppose we have a sequence of simplices satisfying the chain of inequalities. Then for each i, we must have $\{\alpha_i < \beta_i\} \in V$ by definition. Moreover, since the definition of a discrete Morse function forces β_i to be the unique such simplex, each α_i and β_i occur in precisely one such pair. The sequence is thus a V-path. □

This lemma tells us that the gradient paths of a discrete Morse function are those along which f is decreasing.

Theorem 6.19. *A discrete vector field V on X is the gradient vector field of a discrete Morse function if and only if there are no non-trivial closed V-paths.*

Proof. Suppose $V = -\nabla f$ for a discrete Morse function f and suppose we have a non-trivial closed V-path:

$$\alpha_0 < \beta_0 > \alpha_1 < \beta_1 > \cdots < \beta_{r-1} > \alpha_r = \alpha_0.$$

Then by Lemma 6.18 we have the chain of inequalities

$$f(\alpha_0) \geq f(\beta_0) > f(\alpha_1) \geq f(\beta_1) > \cdots \geq f(\beta_{r-1}) > f(\alpha_r) = f(\alpha_0).$$

We therefore cannot have any closed V-paths.

Conversely, suppose V is a discrete vector field with no non-trivial closed V-paths. For notational convenience, if $\{\alpha < \beta\} \in V$, we write $V(\beta) = \alpha$ and $V(\alpha) = \beta$. If σ is a critical cell of V, then we write $V(\sigma) = 0$. For each p, denote by $X^{(p)}$ the p-skeleton of X and define a discrete vector field V_p on $X^{(p)}$ by setting $V_p(\sigma^{(q)}) = V(\sigma^{(q)})$ if $q < p$ and $V(\sigma^{(q)}) = 0$ if $q = p$. We now inductively construct a discrete Morse function f_p on each $X^{(p)}$ whose gradient is V_p and such that if $\sigma^{(q)}$ is critical for f_p then $f_p(\sigma) = q$. Moreover, f_p will have image contained in the closed interval $[-1/2, p + 1/2]$.

Consider $p = 0$. Then $X^{(0)}$ is the 0-skeleton of X and every vertex is critical for V_0. Set $f_0(v) = 0$ for each vertex v. This establishes the base step.

Now suppose we have defined a discrete Morse function f_{p-1} on $X^{(p-1)}$ whose gradient is V_{p-1} and satisfies $f_{p-1}(\sigma^{(q)}) = q$ for every critical cell $\sigma^{(q)}$ for f_{p-1}. We now define the function f_p. If $q \leq p-2$, set $f_p(\sigma^{(q)}) = f_{p-1}(\sigma^{(q)})$. Now consider a $(p-1)$-cell $\sigma^{(p-1)}$ and define $d(\sigma)$ to be

$$d(\sigma) = \max\{r : \text{there is a } V_p\text{-path } \sigma < \tau > \sigma_1 < \tau_1 < \cdots > \sigma_r, V_p(\sigma_r) = 0\}.$$

Each $d(\sigma)$ is finite since V_p has no closed paths (as V has no closed paths). Take D to be the maximum of the $d(\sigma)$ over all the $(p-1)$-cells. We now set

$$f_p(\sigma^{(p-1)}) = f_{p-1}(\sigma^{(p-1)}) + \frac{d(\sigma^{(p-1)})}{2D+1}.$$

Since $f_{p-1}(\sigma) \leq p - 1/2$, we have $f_p(\sigma) < p$. Finally, we define f_p on the p-cells. We have $V_p(\sigma^{(p)}) = 0$. If $\sigma^{(p)}$ is critical for V, we set $f(\sigma) = p$. If $\sigma^{(p)} = V(\alpha^{(p-1)})$, we set $f_p(\sigma) = f_p(\alpha)$. Observe that if $\beta^{(p-1)}$ is another face of $\sigma^{(p)}$, then $f_p(\alpha) > f_p(\beta)$. Indeed, if $\beta < \gamma > \tau_1 < \cdots > \tau_r$ is a V_p-path then $\alpha < \sigma > \beta < \gamma > \tau_1 < \cdots > \tau_r$ is a V_p-path and so $d(\alpha) > d(\beta)$.

We claim that f_p has the requisite properties. First note that the image of f_p is contained in $[-1/2, p + 1/2]$ by construction. If σ is a p-cell then the number of $(p-1)$-cells with f_p-value greater than or equal to that of σ is at most 1 by construction and there are no $(p+1)$-cells in $X^{(p)}$ so that the other condition of Definition 6.3 is satisfied trivially. Now suppose $\ell < p$, σ is an ℓ-cell, and $\tau^{(p-1)} < \sigma$. We claim the following:

$$f_p(\sigma) < f_p(\tau) \Leftrightarrow f_{p-1}(\sigma) < f_{p-1}(\tau).$$

Indeed, $f_p(\tau) = f_{p-1}(\tau)$ since $\dim \tau \leq p-2$. If $\ell < p-1$ then the same is true of τ so that the equivalence holds. If $\ell = p-1$, then $f(\sigma) > f_{p-1}(\sigma)$ and so if $f_{p-1}(\sigma) > f_{p-1}(\tau)$ the equivalence holds. Finally if $f_{p-1}(\sigma) < f_{p-1}(\tau)$, then $V_p(\tau) = \sigma$. But then $d(\sigma) = 0$ and $f_p(\sigma) = f_{p-1}(\sigma)$ and the equivalence holds. In particular, this equivalence implies, by induction, the first condition of Definition 6.3 for $\dim \sigma \leq p-2$ and the second condition for $\dim \sigma \leq p-1$. If $\dim \sigma = p-1$, then, by construction, if $\tau^{(p+1)} > \sigma$ we have $V_p(\sigma) = \tau$. This implies the first condition of Definition 6.3 in this case and so f_p is a discrete Morse function on $X^{(p)}$.

It remains to check that $V_p = -\nabla f_p$. We must show that if $\alpha^{(\ell-1)} < \beta^{(\ell)}$ then

$$f_p(\alpha) > f_p(\beta) \Leftrightarrow V_p(\alpha) = \beta.$$

By induction, this is true for $\ell \leq p-1$ and for $\ell = p$ it is true by construction.

Finally, we claim that if $\sigma^{(q)}$ is critical for f_p, then $f_p(\sigma) = q$. For $q = p$ it is true by construction. If $q < p-1$ then $f_p(\sigma) = f_{p-1}(\sigma)$ and this equals q by induction. If $q = p-1$, then if σ is critical we have $d(\sigma) = 0$, which implies $f_p(\sigma) = f_{p-1}(\sigma)$ and since σ is critical for f_{p-1} the result follows by induction. This completes the proof. □

In our study of smooth Morse theory, we often made use of the fact that if $f : M \to \mathbb{R}$ is a Morse function, then so is $-f$. Moreover, if x is an index-p critical point of f, x is an index-$(m-p)$ critical point of $-f$. This allowed us to prove Poincaré duality, for example. In the discrete case, if f is a discrete Morse function on a complex X, the function $-f$ is *not* a discrete Morse function. However, using the associated gradient vector field, we can define an object that plays the role of $-f$. For simplicity, we will restrict attention to a certain class of regular cell complexes.

Suppose X is a triangulated m-manifold and V is a discrete vector field on X. Recall that the dual complex X^* is defined as follows. If σ is a simplex in X, let τ be an m-simplex containing σ. The vertices of σ then form a subset of the vertices of τ. Denote by σ^* the dual cell of σ defined by the condition that $\tau \cap \sigma^*$ is the convex hull (in τ) of the barycenters of all subsets of the vertices of τ that contain the vertices of σ. It is easy to see that if σ is an i-simplex, then σ^* is an $(m-i)$-simplex. Note that $\sigma \cap \sigma^*$ consists of a single point. It is useful to compare this construction to the dual handlebody decomposition of a manifold induced by the negative of a Morse function on a manifold.

Definition 6.20. Let X be a triangulated m-manifold with a discrete vector field V. The vector field $-V$ is defined on the dual triangulation X^* by the condition

$$\{\alpha < \beta\} \in V \Leftrightarrow \{\beta^* < \alpha^*\} \in -V.$$

Note that a p-simplex σ is critical for V if and only if σ^* is a critical $(m-p)$-simplex for $-V$. Thus we have a duality

$$n_p(V) = n_{m-p}(-V),$$

where $n_p(V)$ denotes the number of critical simplices of index p for V. An example of a discrete vector field V and its dual $-V$ on the boundary of a tetrahedron is shown in Figure 6.7. In the figure, we have drawn the dual complex embedded inside the original tetrahedron to improve the visualization; the true dual triangulation of the original space is obtained by projecting the positive dimensional simplices onto the tetrahedron. The critical cells for V are the top vertex, back and bottom sides,

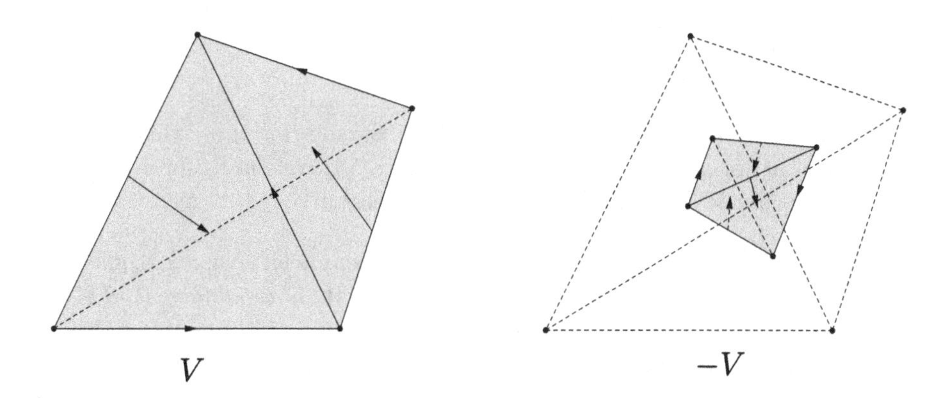

Fig. 6.7 A discrete vector field and its dual

and the edge adjacent to those critical sides. For $-V$, the critical cells are the duals: the top face, the top and bottom vertices, and the edge joining those two vertices.

This is a bit unsatisfying since we are forced to use different triangulations of the space X. However, if we are willing to pass to a barycentric subdivision then we can eliminate this. Indeed, we have $\text{sd}(X) = \text{sd}(X^*)$, so we need only address the following question: given V, can we find a discrete vector field W on $\text{sd}(X)$ which "refines" V in a suitable sense? This was studied in [King et al. (2014)]; the results are as follows.

Definition 6.21. Let V be a discrete vector field on a simplicial complex X and let Y be a refinement of X (e.g., a barycentric subdivision). If σ is a cell of Y let $g(\sigma)$ denote the smallest dimensional cell of X so that $\sigma \subset g(\sigma)$. We say that a discrete vector field W on Y is a *refinement* of V if for each k-cell α of X we can choose a k-cell $h(\alpha)$ of Y so that:

(1) $h(\alpha) \subset \alpha$.
(2) If α is critical then $h(\alpha)$ is critical.
(3) If $\{\alpha, \beta\} \in V$ then $\{h(\alpha), h(\beta)\} \in W$.
(4) If $\{\kappa, \sigma\} \in W$ and $\kappa \neq hg(\kappa)$ then $g(\kappa) = g(\sigma)$.

Definition 6.22. Let V be a discrete vector field on a simplicial complex X and let W be a discrete vector field on a refinement Y of X. If σ is a cell of N let $g(\sigma)$ denote the smallest dimensional cell of X so that $\sigma \subset g(\sigma)$. We say that a W-path $\kappa_0^{(p)}, \sigma_0^{(p+1)}, \kappa_1^{(p)}, \sigma_1^{(p+1)}, \ldots, \sigma_s^{(p+1)}, \kappa_{s+1}^{(p)}$ in Y is a *refinement* of a V-path $\alpha_0^{(p)}, \beta_0^{(p+1)}, \alpha_1^{(p)}, \beta_1^{(p+1)}, \ldots, \beta_r^{(p+1)}, \alpha_{r+1}^{(p)}$ in X if there are indices $0 = i_0 < i_1 < \cdots < i_{r+1} = s + 1$ so that

(1) $g(\kappa_{i_\ell}) = \alpha_\ell$, $0 < \ell \leq r$.
(2) $g(\kappa_i) = \beta_\ell$ if $i_\ell < i < i_{\ell+1}$.

(3) $g(\sigma_i) = \beta_\ell$ if $i_\ell \leq i < i_{\ell+1}$.
(4) If $g(\kappa_0) \neq \alpha_0$ then $g(\kappa_0) = \beta_0$ and α_0 is absent.
(5) If $g(\kappa_{s+1}) \neq \alpha_{r+1}$ then $g(\kappa_{s+1}) = \beta_r$ and α_{r+1} is absent.

In other words, the W-path (optionally) starts in the interior of α_0, then travels in the interior of β_0 for a while, then passes through α_1, travels in β_1 for a while, and so forth. Note that W is *not* necessarily a refinement of V.

Lemma 6.23. *Let V be a discrete vector field on a simplicial complex X and let W be a discrete vector field on a refinement Y of X. If W is a refinement of V then every W-path is a concatenation of two paths: first (optionally) a path contained in a single cell α of X followed (optionally) by a refinement of a V-path. If the initial path in α is needed and the cells of the path have dimensions p and $p+1$, then the initial cell in the W-path is not contained in any cell of X with dimension $< p+2$.*

Proof. Take any W-path $\kappa_0^{(p)}, \sigma_0^{(p+1)}, \kappa_1^{(p)}, \sigma_1^{(p+1)}, \ldots, \sigma_s^{(p+1)}, \kappa_{s+1}^{(p)}$ in N. Let g and h be as in the definition of the refinement of a discrete vector field. Let $\tau_i = g(\kappa_i)$ and $\alpha = \tau_0$. Note that if $\kappa_i \neq h(\tau_i)$ then $g(\sigma_i) = \tau_i$; thus, $\tau_{i+1} = g(\kappa_{i+1}) \subset g(\sigma_i) = \tau_i$. It follows that if the path is not completely contained in α there must be a lowest index j so that $\kappa_j = h(\tau_j)$. Since $\kappa_i \subset \alpha$ and $\sigma_i \subset \alpha$ for all $i < j$ we may as well suppose that $j = 0$ and prove that in this case, the W-path is a refinement of a V-path. Let $\alpha_0 = \tau_0$. Then there is a unique $(p+1)$-cell β_0 so that $(\alpha_0, \beta_0) \in V$ and $h(\beta_0) = \sigma_0$. We define the α_j, β_j and i_j inductively as follows.

- Let $i_0 = 0$, $\alpha_0 = \tau_0$ and $(\alpha_0, \beta_0) \in V$. Since W is a refinement of V, it follows that $\kappa_0 = h(\tau_0)$ (and thus $h(\tau_0) \neq \beta_0$) and $\sigma_0 = h(\beta_0)$.
- Assume that i_{j-1}, α_{j-1} and β_{j-1} are such that $\kappa_{i_{j-1}} = h(\tau_{i_{j-1}})$, $\alpha_{j-1} = \tau_{i_{j-1}}$ and $(\alpha_{j-1}, \beta_{j-1}) \in V$. Let i_j be the least index $> i_{j-1}$ such that $h(\tau_{i_j}) \neq \beta_{i_{j-1}}$.
  - If $i_j = s+1$, terminate (with $r = j-1$),
  - if $i_j < s+1$, then $g(\sigma_{i_j}) \neq g(\kappa_{i_j})$, so let $\alpha_j = g(\kappa_{i_j})$ and β_j the pair of α_j in V, that is, $(\alpha_j, \beta_j) \in V$,
  - if such an index does not exist, terminate (with $r = j-1$ and α_{r+1} absent).

If $i_\ell < i < i_{\ell+1}$ then $\tau_i = \beta_\ell$ and for dimension reasons, $\kappa_i \neq h(\tau_i)$. It follows that $\beta_\ell = g(\kappa_i) = g(\sigma_i)$.

Suppose $\dim \tau_0 < p+2$. We must show that the whole path is a refinement of a V-path. Note that $\dim \tau_0 \geq \dim \kappa_0 = p$ so τ_0 has dimension p or $p+1$. If $\dim \tau_0 = p$, then $g(\sigma_0) \neq g(\kappa_0)$. Thus, $\kappa_0 = h(\tau_0)$ and we showed above that the whole path is a refinement of a V-path. If $\dim \tau_0 = p+1$, then $\kappa_0 \neq h(\tau_0)$, and so $g(\sigma_0) = g(\kappa_0) = \tau_0$. We then eliminate α_0 and start out the V-path with $\beta_0 = \tau_0$. Proceeding as above the whole path is a refinement of a V-path. □

Proposition 6.24. *Let V be a discrete vector field on a simplicial complex X. Let X' be obtained from X by a stellar subdivision of the cell κ (see Appendix B). If κ*

is critical, choose one of the new cells κ' of X' with the same dimension as κ. Then there is a discrete vector field V' on X' so that

(1) V' is a refinement of V.
(2) V' agrees with V on the complement of κ, i.e., if $\{\alpha, \beta\} \in V$ and $\alpha \neq \kappa$ and $\beta \neq \kappa$ then $\{\alpha, \beta\} \in V'$.
(3) The critical simplices of V' and V are the same except that if κ is critical then κ' replaces κ.
(4) If V is a discrete gradient vector field then V' is a discrete gradient vector field.

Proof. We form X' by replacing the cell κ with the cone $c(\kappa)$. Choose a discrete gradient vector field W on $\partial\kappa$ with only two critical cells (this is possible; see Section 7.2 below). Let τ_1 be the top dimensional critical cell and let τ_0 be the other cell (which has dimension 0). We may assume that if κ is critical then κ' is the cone on τ_1. If κ is not critical, but paired to a smaller dimensional cell then we may assume it is paired to τ_1, i.e., $\{\tau_1, \kappa\} \in V$.

If $\{\alpha, \beta\} \in W$ we put $\{c(\alpha), c(\beta)\}$ in V'. We also put $\{v, c(\tau_0)\}$ in V' where v is the vertex of the cone. If κ is critical then we put all of V in V'. If $\{\tau_1, \kappa\} \in V$ we put $\{\tau_1, c(\tau_1)\}$ in V' and we also put all of V except $\{\tau_1, \kappa\}$ in V'. Finally, if κ is paired with a higher dimensional cell σ, we put $\{c(\tau_1), \sigma\}$ in V' as well as all of V except $\{\kappa, \sigma\}$.

We claim that V' is a refinement of V. If α is a cell of V we put $h(\alpha) = \alpha$ if $\alpha \neq \kappa$ and $h(\kappa) = c(\tau_1)$. The conditions for refinement are then clear.

Now suppose that V is a discrete gradient vector field. Suppose that V' is not, so that there is a nontrivial closed V'-path. By Lemma 6.23 this path is a concatenation of a path contained in a single cell of X and a refinement of a V-path. But the only nontrivial V'-paths contained in a single cell of X must be contained in κ. So our path is a concatenation of a path contained in κ and a refinement of a V-path. Our path cannot just be a refinement of a V-path since that V-path would need to be closed. It follows that a portion of the path lies in κ. By Lemma 6.23 we know that all cells in the path have dimension less than $\dim \kappa$. Note it is not possible for the path to pass through the boundary of κ into κ, since the route to do this is $\tau_1, c(\tau_1)$ which violates the dimension restriction. Hence the closed V'-path must be entirely contained in κ and be disjoint from the boundary of κ. But then by definition of V' a V'-path in κ just consists of the cone on a path in the boundary of κ. Since there are no closed paths in the boundary of κ this cannot happen and V' is therefore a discrete gradient vector field. □

Iterating the construction given in the proof of Proposition 6.24, we see that we have the following result.

Proposition 6.25. *Let V be a discrete gradient vector field on a triangulated m-manifold X with c_i critical cells of dimension i. Then there is a refinement V' on* sd(X) *with* c_i *critical cells of dimension* i *and*

$$c_i(V') = c_{m-i}(-V'),$$

where $-V'$ *is the dual of* V' *on* $\mathrm{sd}(X^*) = \mathrm{sd}(X)$. □

Observe that Proposition 6.25 gives us a form of Poincaré duality for discrete Morse theory: given a collection $\{c_i : i = 0, \ldots, m = \dim X\}$ of numbers of critical cells for a discrete Morse function on X, there is a discrete Morse function on sd(X) with numbers of critical cells $\{c_{m-i} : i = 0, \ldots, m = \dim X\}$.

Remark 6.26. Benedetti [Benedetti (2012)] proved a version of Proposition 6.25 for PL manifolds with boundary. There are some technical difficulties to be overcome in defining the dual complex.

6.4 Dynamics of discrete vector fields

Although it is not our main concern, we pause here to discuss the dynamics associated to an arbitrary discrete vector field. Gradient fields have no closed V-paths, but an arbitrary discrete vector field may not have this property. For example, the vector field shown in Figure 6.6 has a loop. We can deduce information about the topology of the cell complex from studying various sets associated to the vector field.

Definition 6.27. Let V be a discrete vector field on the complex X. The *chain recurrent set of* V is the set of cells of X which are either critical for V or are contained in a non-trivial closed V-path.

Definition 6.28. Let V be a discrete vector field on X. Two cells α and β are in the same *basic set* for V if there is a non-trivial closed path which contains both α and β. Also, if σ is a critical cell for V, then the set $\{\sigma\}$ is also a basic set for V.

Example 6.29. Consider the vector field V in Figure 6.6. The chain recurrent set is the entire set of cells of X. There are two basic sets: the critical 2-simplex and the boundary consisting of the three vertices and the three edges.

Now, if Λ is a basic set, define

$$\overline{\Lambda} = \bigcup_{\alpha\in\Lambda} \bigcup_{\beta\subset\overline{\alpha}} \beta.$$

Observe the following. If $\Lambda = \{\sigma\}$ for some critical cell $\sigma^{(p)}$, then

$$H_i(\overline{\Lambda}, \overline{\Lambda} - \Lambda; k) = \begin{cases} k & i = p \\ 0 & i \neq p. \end{cases}$$

If Λ is a union of closed loops, then

$$H_i(\overline{\Lambda}, \overline{\Lambda} - \Lambda; k) = 0, \ i \neq p, p+1.$$

Now define the *Morse numbers* m_i of V by

$$m_i = \sum_{\text{basic sets } \Lambda} \dim H_i(\overline{\Lambda}, \overline{\Lambda} - \Lambda; k).$$

Example 6.30. If V is a discrete gradient vector field, then all the basic sets consist of the critical cells of V. It follows that m_i is simply the number of critical i-cells.

Definition 6.31. Let V be a discrete vector field on X. Denote by K the set of cells of X. A *Lyapunov function* for V is a function $f : K \to \mathbb{R}$ such that for any V-path $\alpha_0 < \beta_0 > \alpha_1$, we have $f(\alpha_0) \geq f(\beta_0) \geq f(\alpha_1)$ and $f(\alpha_0) = f(\alpha_1)$ if and only if α_0 and α_1 are in the same basic set (and hence so is β_0).

Example 6.32. If V is a discrete gradient then a Lyapunov function for V is a discrete Morse function on X.

Proposition 6.33. *Let V be a discrete gradient vector field on X. Then there is a Lyapunov function for V.*

Proof. The proof is similar to the proof of Theorem 6.19. A Lyapunov function is constant on each basic set and if we mimic the proof of Theorem 6.19 we may construct a Lyapunov function which takes the value i on a basic set consisting of $(i-1)$ and i-cells (or a critical cell of index i). The details are left as an exercise. $\square$

6.5 A graph-theoretic point of view

By now, the reader familiar with the theory of directed graphs will have noticed similarities between the notion of a discrete vector field on a complex X and acyclic matchings on a directed graph. This connection was first noticed by Chari [Chari (2000)].

Definition 6.34. Let X be a regular CW-complex. The *Hasse diagram* of X is the directed graph whose vertices are the cells in X and edges are the directed edges from β to α for $\alpha^{(p)} < \beta^{(p+1)}$.

Example 6.35. Consider the sphere S^2 with the CW-decomposition given by taking two cells in each dimension $i = 0, 1, 2$. That is, we have two vertices, v_0, v_1 (the north and south poles), two edges e_0, e_1 (two halves of a great circle through the poles), and two 2-cells f_0, f_1 (the two hemispheres). The Hasse diagram of this decomposition is shown in Figure 6.8.

Now suppose we have a discrete vector field V on X. In the previous section we visualized this by drawing an arrow from α to β for every pair $\{\alpha^{(p)} < \beta^{(p+1)}\} \in V$. We modify the Hasse diagram of X analogously: if $\{\alpha^{(p)} < \beta^{(p+1)}\} \in V$, reverse the

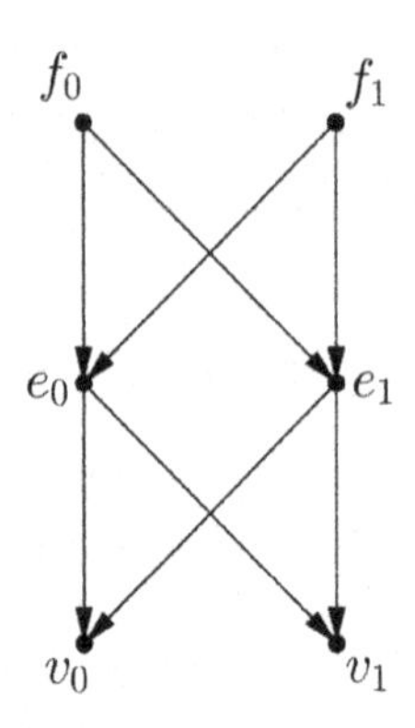

Fig. 6.8 The Hasse diagram of S^2

arrow from β to α in the Hasse diagram. In terms of the modified Hasse diagram, Theorem 6.19 takes the following form.

Theorem 6.36. *A discrete vector field V is the gradent field of a discrete Morse function on X if and only if the modified Hasse diagram has no directed loops.* □

In fact, this characterization may be proved using the following well-known fact from graph theory [Bang-Jensen and Gutin (2008)] in conjunction with Lemma 6.18: A directed graph G has no directed loops if and only if there is a function $f : \text{Vert}(G) \to \mathbb{R}$ which is strictly decreasing along each directed path.

Definition 6.37. Let G be a directed graph. A *partial matching* $\mathcal{M}$ in G is a subset of the edges of G such that each vertex is adjacent to at most one edge in $\mathcal{M}$. The partial matching $\mathcal{M}$ is *acyclic* if upon reversing the orientation of the edges in $\mathcal{M}$, the resulting graph has no directed loops.

In these terms, then, a discrete vector field on X is a partial matching on the Hasse diagram of X, and it is a gradient vector field if the matching is acyclic. In Figure 6.9 we present an acyclic partial matching on the Hasse diagram of the sphere S^2. The edges in the matching are show in gray, with their orientations reversed. Denote by V the resulting gradient vector field and note that the cells v_0 and f_0 are critical for V.

The connection between discrete vector fields and directed graphs is significant. It turns questions about discrete Morse theory into problems in graph theory, where there is a large encyclopedia of algorithms available. We shall exploit this relationship in future chapters.

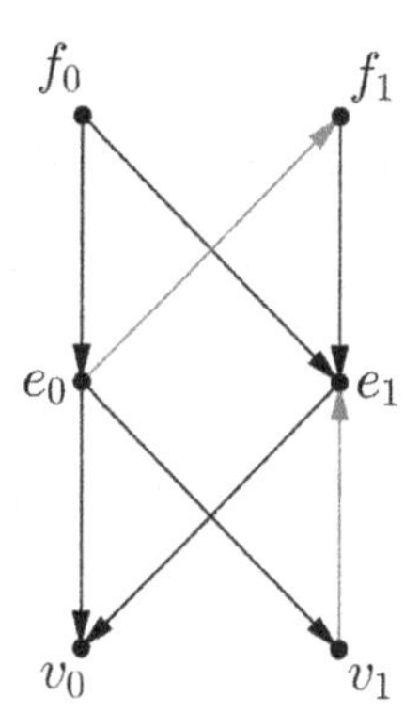

Fig. 6.9 A partial acyclic matching on S^2

6.6 Exercises

(1) Consider the discrete Morse function in Example 6.7. Describe the critical cells.
(2) Extend the discrete Morse function of Figure 6.1 to a discrete Morse function on the 2-simplex in two ways: one in which the 2-cell is critical and one in which it is not.
(3) Locate the critical cells for the discrete Morse function shown in Figure 6.2.
(4) Suppose X is a regular CW-complex. Show that the minimum of a discrete Morse function on X must occur at a vertex.
(5) Suppose M is a triangulated n-manifold. Show that the maximum of a discrete Morse function on M must occur on an n-simplex.
(6) Complete the proof of Lemma 6.11 by proving the following result. In a CW complex X, if $\tau^{(p+1)} > \sigma^{(p)} > \alpha^{(p-1)}$, then one of the following is true.

    (a) σ is an irregular face of τ.
    (b) α is an irregular face of σ.
    (c) There is a p-cell $\mu \neq \sigma$ with $\tau > \mu > \alpha$.

    (Hint: assume the first two conditions are true and use the boundary map in the chain complex $C_\bullet(X)$ to deduce the existence of μ.)
(7) Is the discrete vector field shown in Figure 6.6 the gradient of a discrete Morse function?
(8) Consider the vector field V on the tetrahedron shown in Figure 6.7. Construct a refinement V' of V on $\mathrm{sd}(V)$ and show that the dual $-V'$ satisfies $c_i(V') = c_{m-i}(-V)$.
(9) Prove Proposition 6.33.

Bibliographic notes

The basic results about discrete Morse theory in this chapter are due to Forman and the proofs presented here follow those of his original paper [Forman (1998a)]. The reader interested in more general discrete vector fields, as mentioned in Section 6.4, is referred to Forman's paper [Forman (1998b)]. There is one other characterization of discrete Morse functions as "poset maps with small fibers" due to Kozlov; we refer the reader to his excellent book [Kozlov (2008)] for further details.

Chapter 7

Topological Consequences

If $f : M \to \mathbb{R}$ is a Morse function on a smooth manifold, we are able to deduce the homotopy type of M using the critical points of f. In particular, we had the following results: if $[a, b]$ contains no critical values of f, then M_a is a deformation retract of M_b (Theorem 3.1); and if there is a single critical value in $[a, b]$, then M_b is obtained from M_a by attaching an i-handle, where i is the index of the critical point in the interval (Theorem 3.3). The proofs of these results are difficult and technical.

In the discrete case, however, things are a bit simpler. The analogous results hold and we will prove them now.

7.1 Homotopy type

Definition 7.1. Suppose X is a cell complex and that f is a discrete Morse function on X. Denote by K the set of cells of X. If $a \in \mathbb{R}$, the *sublevel complex* is the cell complex

$$X(a) = \bigcup_{\substack{\sigma \in K \\ f(\sigma) \leq a}} \bigcup_{\tau \leq \sigma} \tau.$$

Theorem 7.2. *Let X be a regular CW complex and suppose f is a discrete Morse function on X. If f has no critical values in $[a, b]$, then*

$$X(b) \searrow X(a).$$

Proof. First note that we may assume that $f : K \to \mathbb{R}$ is injective. Indeed, if $f(\tau^{(p+1)}) \leq f(\sigma^{(p)})$, then we may perturb f by decreasing $f(\tau)$ or increasing $f(\sigma)$ by a small ε without altering the critical cells. Moroever, if $\tau > \sigma > \mu$, all with distinct f values, then we may perturb $f(\sigma)$ without altering the critical cells. Thus, we can always perturb f slightly without changing the sublevel complexes $X(a)$ and $X(b)$ to make f injective.

Note also that we may assume that there is a single noncritical cell in $f^{-1}([a,b])$. By Lemma 6.11, one of the following is true:

(1) There is a $\tau^{(p+1)} > \sigma$ with $f(\tau) < f(\sigma)$; or
(2) There is a $\mu^{(p-1)} < \sigma$ with $f(\mu) > f(\sigma)$.

In the first case, we must have $f(\tau) < a$ and hence $\tau \subseteq X(a)$. Since σ is a face of τ, we then have $\sigma \subseteq X(a)$ as well. Thus, $X(a) = X(b)$ and there is nothing to prove. In the second case, since the first case is not true we must have $f(\tau) > f(\sigma)$ for all $\tau^{(p+1)} > \sigma$. By Lemma 6.12 it follows that $f(\tau) > b$ for any such τ and thus $\sigma \cap X(a) = \emptyset$. Now, we have assumed that there is a $\mu < \sigma$ with $f(\mu) > f(\sigma)$, and so $f(\mu) > b$. If η is any other $(p-1)$-face of σ, then by definition we must have $f(\eta) < f(\sigma)$ and hence that $f(\eta) < a$. Thus, η and all its faces are contained in $X(a)$.

Now, if $\tilde{\sigma} \neq \sigma$ is another p-cell of X containing μ, then we must have $f(\tilde{\sigma}) > f(\mu) > b$. It follows from Lemma 6.12 that if $\tilde{\sigma}$ is any cell of any dimension containing μ then $f(\tilde{\sigma}) > b$ and hence $\mu \cap X(a) = \emptyset$. We therefore conclude that

$$X(b) = X(a) \cup \sigma \cup \mu,$$

where μ is a free face of σ. It follows that $X(b) \searrow X(a)$. □

We also have the following analogue of Theorem 3.3.

Theorem 7.3. *Suppose X is a regular CW complex and f is a discrete Morse function on X. Suppose further that $\sigma^{(p)}$ is a critical cell for f, $f(\sigma) \in [a,b]$, and there are no other critical cells in $f^{-1}([a,b])$. Then $X(b)$ is homotopy equivalent to*

$$X(a) \cup_{\partial e^p} e^p.$$

Proof. Note that we may assume that f is injective and that the only cell in $f^{-1}([a,b])$ is σ (we may need to shrink the interval $[a,b]$, but since there are no other critical cells in this interval Theorem 7.2 guarantees that the homotopy type is unchanged). Since σ is critical, for each $\tau^{(p+1)} > \sigma$ we have $f(\tau) > f(\sigma)$. It follows that $f(\tau) > b$ and by Lemma 6.12 the same is true for any cell $\eta > \sigma$. Thus, $\sigma \cap X(a) = \emptyset$. Also, if $\nu^{(p-1)} < \sigma$, then $f(\nu) < f(\sigma)$ and hence $f(\nu) < a$. It follows that $\nu \subseteq X(a)$ and hence $\partial\sigma \subseteq X(a)$. But then

$$X(b) = X(a) \cup_{\partial\sigma} \sigma$$

and since σ is homeomorphic to a p-cell e^p we are done. □

Contrast the proofs of Theorems 7.2 and 7.3 with the smooth versions. They are significantly shorter and simpler, especially Theorem 7.3. The smooth proofs required a careful analysis of the gradient-like vector fields for the Morse function near a critical point, and this is much trickier than the discrete case, which comes down to simple combinatorial arguments. Of course, Theorems 3.1 and 3.3 concern the *homeomorphism* type of a manifold, so we should expect more complexity, but from the point of view of computing homology, for example, the discrete version is sufficient and much more efficient.

Combining these two results we have the following.

Theorem 7.4. *Let X be a regular CW complex and let f be a discrete Morse function on X. Then X is homotopy equivalent to a CW complex having one cell of index i for each critical cell of index i for f.* □

Observe that Theorem 7.4 immediately implies the following result.

Corollary 7.5. (Weak Morse Inequalities) *Suppose X is a regular CW complex and let f be a discrete Morse function on X. Denote by m_p the number of critical cells of dimension p for f. Then for each p,*

$$m_p \geq \beta_p(X).$$

Moreover,

$$\chi(X) = \sum_{i=0}^{\dim X} (-1)^i m_i.$$

□

The strong Morse inequalities (cf. Theorem 4.20) also follow from Theorem 7.3, of course, but the proof takes a bit more work.

7.2 Sphere theorems

Recall Theorem 2.6: If M is a compact n-manifold supporting a Morse function with exactly two critical points, then M is homeomorphic to S^n. In this section, we show the same is true in the discrete case. We first note some technical results.

Lemma 7.6. *Let X be a CW complex and let Y be a subcomplex. If f is a discrete Morse function on X, then the restriction of f to Y is a discrete Morse function on Y. Moreover, if $\sigma \subseteq Y$ is critical for f, then σ is critical for the restriction as well.*

Proof. This is obvious from the definitions. □

Lemma 7.7. *Let X be a CW complex and Y a subcomplex. If f is discrete Morse function on Y, then f extends to a discrete Morse function $\tilde{f}$ on X.*

Proof. Let c be the maximum value of f on Y. Then if τ is a cell in $X - Y$, set $\tilde{f}(\tau) = c + \dim \tau$. If σ is a cell in Y, set $\tilde{f}(\sigma) = f(\sigma)$. Then $\tilde{f}$ is a discrete Morse function on X. □

Note that the extension $\tilde{f}$ defined in the proof of Lemma 7.7 is not very efficient. Indeed, every cell in $X-Y$ is critical for $\tilde{f}$. One would really prefer an extension with relatively few additional critical cells. Of course we should expect some additional critical cells in general since X will typically have a different homotopy type than Y. However, if X collapses to Y, then we can say something.

Lemma 7.8. *Suppose Y is a subcomplex of X with $X \searrow Y$. Let f be a discrete Morse function on Y and let c be the maximum value of f on Y. Then f may be extended to a discrete Morse function $\tilde{f}$ on X such that $Y = X(c)$ and $\tilde{f}$ has no critical points in $X - Y$.*

Proof. Suppose X collapses to Y via a single collapse: σ is a cell in X with a free face τ and $X = Y \cup \sigma \cup \tau$. We then define $\tilde{f}$ by setting $\tilde{f} = f$ on Y, $\tilde{f}(\sigma) = c+1$, and $\tilde{f}(\tau) = c+2$. Then $\tilde{f}$ is a discrete Morse function with no critical cells in $X-Y$. The result then follows by induction on the number of collapses required. □

As a consequence of Lemma 7.8 we see that if Δ^n is the standard simplex, then Δ^n supports a discrete Morse function with exactly one critical point. Indeed, if v is any vertex, then $\Delta^n \searrow v$. Moreover, the simplicial $(n-1)$-sphere $\partial\Delta^n$ supports a discrete Morse function with exactly two critical points. This follows from the fact that if σ is any $(n-1)$-simplex of $\partial\Delta^n$, the complex $\partial\Delta^n - \sigma$ collapses onto any vertex. The construction of such functions is left as an exercise, but note that it is much more complicated to build a discrete Morse function on $\partial\Delta^n$ than it is to find a smooth Morse function on the sphere S^{n-1} (the standard height function works in this case).

We now prove the discrete analogue of Theorem 2.6.

Theorem 7.9. *Suppose X is a regular CW complex and let f be a discrete Morse function on X with exactly two critical cells. Then X is homotopy equivalent to a sphere. If, in addition, X is a compact PL n-manifold without boundary, then X is piecewise linear equivalent to $\partial\Delta^{n+1}$ with its standard triangulation.*

Proof. Since X is connected, the weak Morse inequalities imply that at least one critical cell of f has index 0. If the other critical cell has index n, then by Theorem 7.3, X is homotopy equivalent to an n-cell with its boundary collapsed to a point; that is, $X \simeq S^n$. If, in addition, X is a compact PL n-manifold, then $H_n(X, \mathbb{Z}/2) \neq 0$, which implies that there is a critical cell of index n, say σ. Let $Y = X - \sigma$. Then f, restricted to Y, has a single critical cell occurring at a vertex v. By Theorem 7.2, $Y \searrow v$. By Whitehead's Theorem [Stallings (1968)], Y is PL n-cell. Since $X = Y \cup_{\partial\sigma} \overline{\sigma}$ and $\overline{\sigma}$ is clearly a PL n-cell, it follows that X is a PL n-sphere. □

Remark 7.10. It is possible to say more about Theorem 7.9. Assuming X is a polyhedron that is topologically a manifold without boundary, the existence of a discrete Morse function on X with two critical cells implies that X is *homeomorphic* to S^n ([Forman (1998a)]). The proof requires the Poincaré Conjecture, however.

Recall Theorem 3.22, which asserts that a connected closed m-manifold M supports a Morse function with exactly one critical point of index 0 and one critical point of index m. There is a discrete version of this result.

Theorem 7.11. *Suppose X is a triangulated connected closed m-manifold. Then X has a discrete Morse function with a single critical vertex and a single critical m-cell.*

Proof. We first note that if X is any connected polyhedron, with or without boundary, the 1-skeleton $X^{(1)}$ is a connected graph. Let T be a maximal tree in $X^{(1)}$. Then if v is any vertex, $T \searrow v$ and thus there is a discrete Morse function on T such that v is the only critical vertex. By Lemma 7.7, we may extend this to a discrete Morse function f on X. Since $X - T$ contains no vertices, v is the only critical point of f of index 0.

If $\dim X = 0$, then there is nothing more to prove. If $\dim X = 1$, then X is a circle. If e is any edge, Then $X - e \searrow v$, where v is any vertex. It follows that $X - e$ has a discrete Morse function f with the single critical cell, v. Setting c to be the maximum of f on $X - e$, define $f(e) = c + 1$.

Assume $\dim X \geq 2$. Let σ be any m-simplex in X and let $Y = X - \sigma$. Then Y is a triangulated m-manifold with boundary and by collapsing the m-simplices of Y one at a time along the free $(m-1)$-faces, we see that $Y \searrow Z$, where Z is a subcomplex of dimension at most $m - 1$. By the remark above, Z has a discrete Morse function f with a single index 0 critical cell, which we may take to be any vertex v. By Lemma 7.8, f extends to Y with no additional critical cells. Setting $f(\sigma) = c + 1$, where c is the maximum value of f on Y, we obtain the required discrete Morse function on X. □

Theorem 3.22 was proved by first canceling 0- and 1-handles to reduce to a single index 0 critical point and then repeating the procedure for the negative of the resulting Morse function. We could have tried that approach here, but the problem is that we cannot simply negate a discrete Morse function. We have the dual vector field associated to $V = -\nabla f$, but this is on the *dual complex* X^*. This argument will therefore not work in the discrete case.

7.3 Canceling critical cells

We now turn to the question of canceling critical cells of a discrete Morse function. Recall Theorem 3.21, which asserts that if a Morse function f on a manifold M has a pair of critical points p and q of indices differing by 1 whose ascending and descending manifolds intersect transversely, then f may be perturbed to a Morse function g which has the same critical points except that p and q are no longer critical. In other words, we may "cancel" the critical points p and q, thereby obtaining a simpler Morse function.

The proof of Theorem 3.21 is long and technical; indeed, we did not provide a complete proof. It is natural to ask if an analogous result holds in the discrete case, and that is the goal of this section. Note that the essential reason Theorem 3.21 is true is that there is a single gradient path between the critical points p and q (see Figure 3.4 for an illustration). In the discrete case, it makes no real sense to talk about "perturbations" of discrete Morse functions since small changes in the values of such a function typically do not affect anything. The correct point of view is to make alterations to the gradient vector field of the discrete Morse function.

The basic idea of the proof of Theorem 3.21 is illustrated in Figure 3.5. The gradient-like vector field is altered in a neighborhood of the two critical points to eliminate both of them. In essence, we are "turning the vector field around" near the points. In the discrete case, this is exactly the right thing to do.

Theorem 7.12. *Suppose that V is a discrete gradient vector field on a regular CW complex X. Suppose $\tau^{(p+1)}$ and $\sigma^{(p)}$ are critical cells such that there is a face $\nu^{(p)} < \tau$ and a unique V-path*

$$\tau > \nu = \sigma_0 < \tau_0 > \sigma_1 < \tau_1 > \cdots < \tau_r > \sigma_r = \sigma.$$

Then there is a gradient vector field W such that the set critical cells of W is the set of critical cells of V with σ and τ removed. Moreover, $W = V$ except along the unique gradient path from $\partial\tau$ to σ.

Proof. This is almost trivial. We simply turn V around along the path from $\partial\tau$ to σ. Define W as follows:

$$\begin{aligned} W(\alpha) &= V(\alpha) \text{ if } \alpha \notin \{\nu, \tau_0, \sigma_1, \tau_1, \ldots, \tau_r, \sigma\} \\ W(\sigma_i) &= \tau_{i-1},\ i = 1, \ldots, r \\ W(\nu) &= \tau. \end{aligned}$$

We need only show that we have not created any closed W-paths. Any such path would have to contain a p-cell in the unique path from $\partial\tau$ to σ and a p-cell not in the path. Thus, a closed path would have to contain a segment of the form

$$\sigma_i < \eta_0 > \mu_1 < \eta_1 > \cdots < \eta_s > \sigma_j$$

with $s \geq 0$ and $\eta_i, \mu_i \neq \sigma_k, \tau_k$ for all i. Since $W(\eta_i) = V(\eta_i)$ and $W(\mu_i) = V(\mu_i)$ for all i, we then have a V-path

$$\eta_0 > \mu_1 < \eta_1 > \cdots < \eta_s > \sigma_j.$$

If $i \neq 0$, then $\mu_1 \neq \sigma_{i-1}, \sigma_i$ and $\mu_1 < W(\sigma_i) = \tau_{i-1}$; thus we may insert this segment into our path to obtain

$$\nu = \sigma_0 < \tau_0 > \sigma_1 < \cdots > \sigma_{i-1} < \tau_{i-1} > \mu_1 < \eta_1 > \cdots > \sigma_j < \tau_j > \cdots > \sigma_r = \sigma,$$

a second gradient path from $\partial\tau$ to σ. This is a contradiction. If $i = 0$, then $\nu \neq \mu_1 < W(\nu) = \tau$ and we may replace the initial segment of the path with this segment to obtain a second gradient path from $\partial\tau$ to σ, also a contradiction. This completes the proof. □

Remark 7.13. Theorem 7.12 tells us how to "perturb" a discrete Morse function to eliminate a pair of critical cells in the following sense. We begin with a gradient field V, which by Theorem 6.19 has a self-indexing discrete Morse function f (i.e., $V = -\nabla f$). We similarly obtain a self-indexing discrete Morse function g such that $W = -\nabla g$. Following the construction given in the proof of Theorem 6.19, we see immediately that $f = g$ except along the unique path from $\partial\tau$ to σ. Thus, we have "perturbed" f in a neighborhood of the path.

When working with discrete Morse functions, one would like them to have as few critical simplices as possible. Theorem 7.12 suggests that this is "easy." However, counting gradient paths between critical cells is nontrivial and can consume a great deal of time. In fact, constructing optimal discrete Morse functions is a difficult problem in general, as shown by Joswig and Pfetsch [Joswig and Pfetsch (2006)]; see Section 8.4.

7.4 Homology

Theorem 7.4 tells us that we may compute the homology of a CW complex using the critical cells of a discrete Morse function on the complex. Recall the definition of the Morse complex given in Section 4.2. For each $i \geq 0$, we take a free abelian group with basis the critical points of index i for a Morse–Smale function f. The differential has a rather complicated definition built by considering quotient spaces of the cells in the Morse–Smale decomposition of the manifold. Ultimately, one sees that the differential in the Morse complex takes a critical point p of index i to a weighted sum of critical points of index $i-1$, where the coefficient on a particular index $i-1$ critical point q is the number of equivalence classes of gradient paths from p to q, counted with orientations. In this section, we will construct the discrete version of this complex, show that the differential is similarly defined, and show that it may be used to compute the homology of our original CW complex.

Let X be a regular CW complex and suppose that V is a discrete gradient vector field on X; say $V = -\nabla f$ for some discrete Morse function f on X. Choose an orientation for each cell in X and consider the chain complex $C_\bullet(X,\mathbb{Z})$ with basis the oriented cells. We will use V to define a map on chains and also construct a discrete analogue of the flow of V. These will then be used to define the discrete Morse complex.

If σ is a p-cell in X, we write

$$\partial\sigma = \sum_{\alpha^{(p-1)}<\sigma} \varepsilon(\sigma,\alpha)\alpha,$$

where $\varepsilon(\sigma,\alpha)$ is the incidence number of α in the boundary of σ. That is, $\varepsilon(\sigma,\alpha)$ counts the number of times α appears in the boundary of σ, counted with multiplicity. Define an inner product $\langle , \rangle$ on $C_\bullet(X,\mathbb{Z})$ by setting the cells to be orthonormal.

In terms of this inner product, the boundary map takes the form

$$\partial\sigma = \sum_{\alpha<\sigma} \langle \partial\sigma, \alpha \rangle \alpha.$$

We define a map

$$V : C_p(X, \mathbb{Z}) \to C_{p+1}(X, \mathbb{Z})$$

as follows. Let σ be a p-cell with fixed orientation. If there is a $(p+1)$-cell τ with $\{\sigma < \tau\} \in V$, then set

$$V(\sigma) = -\langle \partial\tau, \sigma \rangle \tau.$$

If there is no such τ (e.g., if σ is critical), then set $V(\sigma) = 0$. This extends linearly to a map $C_p(X, \mathbb{Z}) \to C_{p+1}(X, \mathbb{Z})$.

Now, let us define the discrete flow Φ. To motivate this, consider the vertices of X. If v is a critical vertex, then v is fixed by the gradient flow; that is, $\Phi(v) = v$. If v is not critical and $V(v) = \pm e$, then v should flow to the other vertex of e; that is,

$$\Phi(v) = v + \partial(V(v)).$$

In general, we define the discrete gradient flow Φ for an oriented p-cell σ by

$$\Phi(\sigma) = \sigma + \partial V(\sigma) + V(\partial\sigma).$$

This really is the correct thing to do. Indeed, if we think of $V = -\nabla f$, then $V(\sigma)$ is the component of $-\nabla f$ transversal to σ, and the component tangent to σ is determined by $V(\partial\sigma)$. The map Φ then extends linearly to a map $\Phi : C_p(X, \mathbb{Z}) \to C_p(X, \mathbb{Z})$.

We note two properties of the map V, which follow immediately from the definition of a discrete gradient vector field.

(1) If σ is an oriented p-cell, then

$$\#\{\alpha^{(p-1)} : V(\alpha) = \pm\sigma\} \leq 1.$$

(2) If σ is an oriented p-cell, then

$$\sigma \text{ is critical} \Leftrightarrow \sigma \notin \operatorname{im}(V) \text{ and } V(\sigma) = 0.$$

Moreover, we have the following.

Lemma 7.14. $V \circ V = 0$.

Proof. If $V(\alpha) = \pm\sigma$, then there is no cell $\tau > \sigma$ with $\{\sigma < \tau\} \in V$. Thus, $V(\tau) = 0$. □

Proposition 7.15. *The flow map commutes with the boundary:* $\Phi\partial = \partial\Phi$. *Moreover, if* $\sigma_1, \ldots, \sigma_r$ *are the oriented* p*-cells of* X, *and we write*

$$\Phi(\sigma_i) = \sum a_{ij}\sigma_j$$

then we have

(1) For each i, $a_{ii} = 0$ or 1, and $a_{ii} = 0$ if and only if σ_i is critical;
(2) If $i \neq j$ and $a_{ij} \neq 0$, then $f(\sigma_j) < f(\sigma_i)$.

Proof. Since $\Phi = 1 + \partial V + V\partial$, we have

$$\Phi\partial = (1 + \partial V + V\partial)\partial = \partial + \partial V\partial + V\partial^2 = \partial + \partial V\partial$$
$$\partial V = \partial(1 + \partial V + V\partial) = \partial + \partial^2 V + \partial V\partial = \partial + \partial V\partial.$$

If σ is a p-cell, then we know that σ satisfies either (i) σ is critical, (ii) $\pm\sigma \in \mathrm{im}(V)$, or (iii) $V(\sigma) \neq 0$ (and these conditions are exclusive). Suppose σ is critical. Then $V(\sigma) = 0$ and

$$\Phi(\sigma) = \sigma + V(\partial\sigma) = \sigma + \sum_{\alpha<\sigma} \langle\partial\sigma, \alpha\rangle V(\alpha).$$

Since σ is critical, each such α satisfies $f(\alpha) < f(\sigma)$. Moreover, for each α, we have $V(\alpha) = 0$ or $V(\alpha) = \beta^{(p)}$ with $f(\beta) \leq f(\alpha) < f(\sigma)$. We therefore have

$$\Phi(\sigma) = \sigma + \sum a_\beta \beta$$

where $a_\beta \neq 0$ implies that $f(\beta) < f(\sigma)$.

Suppose next that $\pm\sigma \in \mathrm{im}(V) \subseteq \ker(V)$. Then

$$\Phi(\sigma) = \sigma + V(\partial\sigma) = \sigma + \sum_{\alpha<\sigma} \langle\partial\sigma, \alpha\rangle V(\alpha).$$

There is exactly one α with $V(\alpha) = \pm\sigma$ and $\langle\partial\sigma, \alpha\rangle V(\alpha) = -\sigma$. Also, if $\tilde{\alpha}$ is another face of σ, we have $V(\tilde{\alpha}) = 0$ or $V(\tilde{\alpha}) = \tilde{\sigma}$ with $f(\tilde{\sigma}) \leq f(\tilde{\alpha}) < f(\sigma)$. It follows that

$$\Phi(\sigma) = \sum_{\tilde{\sigma}^{(p)}} a_{\tilde{\sigma}} \tilde{\sigma},$$

where $a_{\tilde{\sigma}} \neq 0$ implies $f(\tilde{\sigma}) < f(\sigma)$.

Finally, if $V(\sigma) = -\langle\partial\tau, \sigma\rangle\tau \neq 0$, then

$$\Phi(\sigma) = \sigma + V(\partial\sigma) + \partial(V(\sigma)).$$

Since $V(\sigma) \neq 0$, $\pm\sigma \notin \mathrm{im}(V)$, and hence for each $(p-1)$-face $\alpha < \sigma$, either $V(\alpha) = 0$ or $V(\alpha) = \pm\tilde{\sigma}$ where $f(\tilde{\sigma}) \leq f(\alpha) < f(\sigma)$. Also,

$$\partial(V(\sigma)) = -\langle\partial\tau, \sigma\rangle\partial\tau = -\langle\partial\tau, \sigma\rangle^2\sigma + \sum a_{\tilde{\sigma}}\tilde{\sigma} = -\sigma + \sum b_{\tilde{\sigma}}\tilde{\sigma},$$

where $b_{\tilde{\sigma}} \neq 0$ implies $f(\tilde{\sigma}) \leq f(\tau) < f(\sigma)$. □

We are now ready to define the discrete Morse complex. Denote by $K_p(X, \mathbb{Z})$ the Φ-invariant chains of X:

$$K_p(X, \mathbb{Z}) = \{c \in C_p(X, \mathbb{Z}) : \Phi(c) = c\}.$$

By Proposition 7.15, the boundary map restricts to a boundary map on $K_\bullet(X, \mathbb{Z})$. We claim that the homology of $K_\bullet(X, \mathbb{Z})$ is the homology of X. To do this, we need a map $C_p(X, \mathbb{Z}) \to K_p(X, \mathbb{Z})$ which behaves nicely with respect to the inclusion map $i : K_p(X, \mathbb{Z}) \to C_p(X, \mathbb{Z})$. That is, we need to assign a Φ-invariant chain to an arbitrary chain in X. We first show that iterating the map Φ stabilizes after a finite number of steps.

Lemma 7.16. *Let $c \in K_p(X, \mathbb{Z})$. If*

$$c = \sum_{\sigma} a_\sigma \sigma,$$

let σ^* be any cell which maximizes $\{f(\sigma) : a_\sigma \neq 0\}$. Then σ^* is critical.

Proof. We have

$$c = \Phi(c) = \sum_{\sigma} a_\sigma \Phi(\sigma).$$

Thus,

$$a_{\sigma^*} = \langle c, \sigma^* \rangle = \sum a_\sigma \langle \Phi(\sigma), \sigma^* \rangle.$$

Proposition 7.15 implies that if $\sigma \neq \sigma^*$ and $f(\sigma) \leq f(\sigma^*)$, then $\langle \Phi(\sigma), \sigma^* \rangle = 0$. It follows that $0 \neq a_{\sigma^*} = a_{\sigma^*} \langle \Phi(\sigma^*), \sigma^* \rangle$ and hence that $\langle \Phi(\sigma^*), \sigma^* \rangle \neq 0$. Proposition 7.15 implies that σ^* is critical. □

Proposition 7.17. *For N sufficiently large, $\Phi^N = \Phi^{N+1} = \cdots$.*

Proof. Let σ be a p-cell in X. We proceed by induction on

$$r = \#\{\tilde{\sigma} : f(\tilde{\sigma}) < f(\sigma)\}.$$

If $r = 0$, then by Proposition 7.15 we have $\Phi(\sigma) = \sigma$ or $\Phi(\sigma) = 0$. In either case $\Phi^N(\sigma) = \Phi^{N+1}(\sigma) = \cdots$ once N is at least 1. For the inductive step, suppose first that σ is not critical. Then by Proposition 7.15

$$\Phi(\sigma) = \sum_{f(\tilde{\sigma}) < f(\sigma)} a_{\tilde{\sigma}} \tilde{\sigma}.$$

By induction, there is a sufficiently large N such that $\Phi^N(\tilde{\sigma})$ is Φ-invariant when $f(\tilde{\sigma}) < f(\sigma)$. Then $\Phi^{N+1}(\sigma)$ is Φ-invariant.

If σ is not critical, let $c = V(\partial\sigma)$. Then

$$\Phi^n(\sigma) = \sigma + c + \Phi(c) + \cdots + \Phi^{n-1}(c).$$

Thus, $\Phi^N(\sigma)$ is Φ-invariant if and only if $\Phi^N(c) = 0$ for some N. We know that c is the sum of p-cells $\tilde{\sigma}$ with $f(\tilde{\sigma}) < f(c)$ (Proposition 7.15) and so by induction there is an $\tilde{N}$ such that $\Phi^{\tilde{N}}(c)$ is Φ-invariant. Now, $c \in \text{im}(V)$ and $\text{im}(V)$ is Φ-invariant:

$$\Phi V = (1 + \partial V + V\partial)V = V(1 + \partial V).$$

It follows that $\Phi^{\tilde{N}}(c) \in \text{im}(V)$. Proposition 7.15 tells us that the image of V is orthogonal to the critical cells; thus $\Phi^{\tilde{N}}(c)$ is a Φ-invariant chain orthogonal to the critical cells, and hence by Lemma 7.16, $\Phi^{\tilde{N}}(c) = 0$. Thus, we have found a large enough N with $\Phi^N(c) = 0$ and hence $\Phi^N(\sigma)$ is Φ-invariant. □

Since X is a finite complex, we may find an N such that for every chain c (of any dimension)

$$\Phi^N(c) = \Phi^{N+1}(c) = \cdots .$$

Denote this Φ-invariant chain by $\Phi^\infty(c)$. Then for every $p \geq 0$ we have a map

$$\Phi^\infty : C_p(X, \mathbb{Z}) \to K_p(X, \mathbb{Z}).$$

Observe that $\Phi^\infty \circ i$ is the identity map on $K_p(X, \mathbb{Z})$.

Theorem 7.18. *For each $p \geq 0$, we have an isomorphism*

$$H_p(K_\bullet(X, \mathbb{Z})) \cong H_p(X, \mathbb{Z}).$$

Proof. Since $\Phi^\infty \circ i$ is the identity map on $K_p(X, \mathbb{Z})$, we have

$$\mathrm{id} = \Phi_*^\infty \circ i_*.$$

We must show that the opposite composition $i_* \circ \Phi_*^\infty$ is the identity as well. To this end, we construct a chain homotopy

$$D : K_p(X, \mathbb{Z}) \to K_{p-1}(X, \mathbb{Z})$$

such that

$$\mathrm{id} - i \circ \Phi^\infty = \partial D + D\partial.$$

Since $\Phi^\infty = \Phi^N$ for some large enough N, we have

$$\begin{aligned}
\mathrm{id} - i \circ \Phi^\infty &= \mathrm{id} - \Phi^N = (\mathrm{id} - \Phi)(\mathrm{id} + \Phi + \cdots + \Phi^{N-1}) \\
&= (-\partial V - V\partial)(\mathrm{id} + \Phi + \cdots + \Phi^{N-1}) \\
&= \partial[-V(\mathrm{id} + \Phi + \cdots + \Phi^{N-1})] + [-V(\mathrm{id} + \Phi + \cdots + \Phi^{N-1})]\partial.
\end{aligned}$$

We may therefore take $D = -V(\mathrm{id} + \Phi + \cdots + \Phi^{N-1})$. This completes the proof. □

So far, we have constructed a complex which computes the homology of X from the data associated to a discrete Morse function, but the boundary map remains a bit mysterious. In the smooth case, we used something like invariant chains, and the boundary map was seen to be expressible in terms of the number of paths connecting pairs of critical points for a Morse function. We would like an analogous description of the boundary map in $K_\bullet(X, \mathbb{Z})$; we will produce that now. We will also show that the complex $K_\bullet(X, \mathbb{Z})$ is isomorphic to the span of the critical cells of V.

For each p, let $\mathbb{M}_p(X, \mathbb{Z})$ denote the span of the critical p-cells in $C_p(X, \mathbb{Z})$. The map Φ^∞ restricts to give a map

$$\Phi^\infty : \mathbb{M}_p(X, \mathbb{Z}) \to K_p(X, \mathbb{Z}).$$

Note the following fact: if σ is a critical p-cell, then for any other critical p-cell $\tilde{\sigma} \neq \sigma$ we have

$$\langle \Phi^{\infty}(\sigma), \tilde{\sigma} \rangle = 0.$$

Indeed, we know that $\Phi^{\infty}(\sigma) = \sigma + c$ where $c \in \operatorname{im}(V)$, and since the image of V is orthogonal to $\mathbb{M}_p(X, \mathbb{Z})$ we obtain the result.

Theorem 7.19. *The map*

$$\Phi^{\infty} : \mathbb{M}_p(X, \mathbb{Z}) \to K_p(X, \mathbb{Z})$$

is an isomorphism.

Proof. Suppose $c \in K_p(X, \mathbb{Z})$ and consider the chain

$$\tilde{c} = \sum_{\sigma \text{ critical}} \langle c, \sigma \rangle \sigma.$$

By the remark above, for any critical cell σ, we have

$$\langle \Phi^{\infty}(\tilde{c}), \sigma \rangle = \langle c, \sigma \rangle.$$

Thus, $\Phi^{\infty}(\tilde{c}) - c$ is a Φ-invariant chain such that for any critical cell σ, $\langle \Phi^{\infty}(\tilde{c}) - c, \sigma \rangle = 0$. Then by Lemma 7.16 we have $\Phi^{\infty}(\tilde{c}) - c = 0$ and hence Φ^{∞} is surjective.

Now suppose $\Phi^{\infty}(c) = 0$ for $c \in \mathbb{M}_p(X, \mathbb{Z})$. Then for any critical cell σ we have $\langle \Phi^{\infty}(c), \sigma \rangle = 0$. The remark above then implies that for any critical cell σ we have $\langle c, \sigma \rangle = 0$ so that $c = 0$; that is Φ^{∞} is injective. □

Denote the boundary map in $\mathbb{M}_{\bullet}(X, \mathbb{Z})$ by d. Theorem 7.19 and Proposition 7.15 imply that d satisfies the following:

$$\langle d(c), \sigma \rangle = \langle \partial \Phi^{\infty}(c), \sigma \rangle = \langle \Phi^{\infty} \partial(c), \sigma \rangle$$

for any critical $(p-1)$-cell σ. We now show that the map d is expressible in terms of the number of gradient paths connecting pairs of critical cells, analogous to the smooth case.

Suppose $\sigma \neq \sigma'$ are p-cells in X and that τ is a $(p+1)$-cell with $\sigma < \tau$ and $\sigma' < \tau$. Recall that we have fixed orientations on the cells of X, but we can ask the following question: if we "slide" σ to σ' through τ, does the orientation induced on σ' agree with the fixed one? Put another way, fixing the orientations on σ and τ, we can induce an orientation on σ' via the following formula:

$$\langle \partial \tau, \sigma \rangle \langle \partial \tau, \sigma' \rangle = -1.$$

So, given a gradient path

$$\sigma_0 < \tau_0 > \sigma_1 < \tau_1 > \cdots < \tau_{r-1} > \sigma_r,$$

an orientation on σ_0 induces an orientation on σ_r. Let γ denote the gradient path and set $m(\gamma) = 1$ if the orientation of σ_r agrees with the fixed one and $m(\gamma) = -1$ otherwise. A formula for $m(\gamma)$ is

$$m(\gamma) = \prod_{i=0}^{r-1} \langle \partial \tau_i, \sigma_i \rangle \langle \partial \tau_i, \sigma_{i+1} \rangle.$$

We call this the *multiplicity* of the path γ. Note that if we concatenate two paths γ_1 and γ_2, then $m(\gamma_1 \circ \gamma_2) = m(\gamma_1)m(\gamma_2)$.

Theorem 7.20. *For any p-cells σ, σ' in X, denote by $\Gamma_r(\sigma, \sigma')$ the set of all gradient paths from σ to σ' of length r. Let $\tau^{(p+1)}$ and $\sigma^{(p)}$ be critical cells for V. Then for sufficiently large N,*

$$\langle d\tau, \sigma \rangle = \sum_{\sigma' < \tau} \langle \partial\tau, \sigma' \rangle \sum_{\gamma \in \Gamma_N(\sigma', \sigma)} m(\gamma).$$

That is, the boundary map in $\mathbb{M}_\bullet(X, \mathbb{Z})$ is expressible in terms of the numbers of gradient paths connecting critical cells, counted with multiplicity.

Proof. Let us first define the reduced gradient flow $\tilde{\Phi} : C_p(X, \mathbb{Z}) \to C_p(X, \mathbb{Z})$ by $\tilde{\Phi} = \mathrm{id} + \partial V$. Suppose $\tau^{(p+1)}$ and $\sigma^{(p)}$ are critical cells. We claim that

$$\langle d\tau, \sigma \rangle = \langle \tilde{\Phi}^\infty \partial\tau, \sigma \rangle.$$

To prove this it suffices to show that for each $r \geq 0$, $\langle \tilde{\Phi}^r \partial\tau, \sigma \rangle = \langle \Phi^r \partial\tau, \sigma \rangle$, since for every chain c and every $r \geq 0$, the difference $\Phi^r(c) - \tilde{\Phi}^r(c)$ lies in the image of V, which is orthogonal to $\mathbb{M}_\bullet(X, \mathbb{Z})$. We proceed by induction on r, the case $r = 0$ being trivial. If $r > 0$, then

$$\Phi^r(c) = \Phi(\Phi^{r-1}(c)) = \Phi(\tilde{\Phi}^{r-1}(c) + V(c'))$$

for some chain c' (by induction). But then we have

$$\begin{aligned}
\Phi^r(c) &= \Phi(\tilde{\Phi}^{r-1}(c) + V(c')) \\
&= (\tilde{\Phi} + V\partial)(\tilde{\Phi}^{r-1}(c) + V(c')) \\
&= \tilde{\Phi}^r(c) + \tilde{\Phi}(V(c')) + V\partial\tilde{\Phi}^{r-1}(c) + V\partial V(c') \\
&= \tilde{\Phi}^r(c) + V(c' + \partial\tilde{\Phi}^{r-1}(c) + V\partial V(c')),
\end{aligned}$$

and this proves the claim.

The reduced gradient flow enjoys the following useful property.

Lemma 7.21. *Suppose σ_1 and σ_2 are any two p-cells in X. Then*

$$\langle \tilde{\Phi}\sigma_1, \sigma_2 \rangle = \sum_{\gamma \in \Gamma_1(\sigma_1, \sigma_2)} m(\gamma).$$

Proof. If $V(\sigma_1) = 0$ then

$$\langle \tilde{\Phi}\sigma_1, \sigma_2 \rangle = \langle \sigma_1, \sigma_2 \rangle = \delta_{ij}.$$

The only gradient path of length 1 beginning at σ_1 is the trivial path $\gamma = \sigma_1$; thus, $m(\gamma) = 1$. So, if $\sigma_1 \neq \sigma_2$, $\Gamma_1(\sigma_1, \sigma_2) = \emptyset$, and thus

$$\sum_{\gamma \in \Gamma(\sigma_1, \sigma_2)} m(\gamma) = 0,$$

while if $\sigma_1 = \sigma_2$, $\Gamma_1(\sigma_1, \sigma_2) = \{\gamma\}$ and
$$\sum_{\gamma \in \Gamma(\sigma_1, \sigma_2)} m(\gamma) = 1.$$

Now, if $V(\sigma_1) \neq 0$, then
$$\langle \tilde{\Phi}\sigma_1, \sigma_1 \rangle = \langle \sigma_1, \sigma_1 \rangle + \langle \partial V(\sigma_1), \sigma_1 \rangle = 1 - 1 = 0.$$
But, since $V(\sigma_1) \neq 0$, there is no gradient path of length 1 from σ_1 to σ_1 and so
$$\sum_{\gamma \in \Gamma(\sigma_1, \sigma_1)} m(\gamma) = 0.$$
If $\sigma_1 \neq \sigma_2$, then
$$\langle \tilde{\Phi}\sigma_1, \sigma_2 \rangle = \langle \sigma_1, \sigma_2 \rangle + \langle \partial V(\sigma_1), \sigma_2 \rangle = \langle \partial V(\sigma_1), \sigma_2 \rangle.$$
If σ_2 is not a face of $V(\sigma_1)$, then $\langle \tilde{\Phi}\sigma_1, \sigma_2 \rangle = 0$ and since there is no gradient path of length 1 from σ_1 to σ_2 the formula holds. If σ_2 is a face of $V(\sigma_1)$, there is exactly one gradient path from σ_1 to σ_2 of length 1: $\gamma = \sigma_1 < V(\sigma_1) > \sigma_2$. But then
$$m(\gamma) = -\langle \sigma_1, \partial V(\sigma_1) \rangle \langle \partial V(\sigma_1), \sigma_2 \rangle = \langle \partial V(\sigma_1), \sigma_2 \rangle.$$
□

We now complete the proof of Theorem 7.20. There is a sufficiently large N so that
$$\langle d\tau, \sigma \rangle = \langle \tilde{\Phi}^N \partial\tau, \sigma \rangle.$$
Since
$$\partial\tau = \sum_{\sigma' < \tau} \langle \partial\tau, \sigma' \rangle \sigma',$$
we have
$$\langle d\tau, \sigma \rangle = \sum_{\sigma' < \tau} \langle \partial\tau, \sigma' \rangle \langle \tilde{\Phi}^N \sigma', \sigma \rangle.$$
We now proceed by induction to prove that for all $r \geq 0$,
$$\langle \tilde{\Phi}^r \sigma', \sigma \rangle = \sum_{\gamma \in \Gamma_r(\sigma', \sigma)} m(\gamma).$$
The case $r = 0$ is trivial and $r = 1$ is the content of Lemma 7.21. If $r > 1$, then
$$\begin{aligned}
\langle \tilde{\Phi}^r \sigma', \sigma \rangle &= \langle \tilde{\Phi}(\tilde{\Phi}^{r-1}\sigma'), \sigma \rangle \\
&= \sum_{\beta^{(p)}} \langle \tilde{\Phi}^{r-1}\sigma', \beta \rangle \langle \tilde{\Phi}\beta, \sigma \rangle \\
&= \sum_{\beta} \sum_{\gamma \in \Gamma_{r-1}(\sigma', \beta)} m(\gamma) \langle \tilde{\Phi}\beta, \sigma \rangle \\
&= \sum_{\beta} \sum_{\gamma \in \Gamma_{r-1}(\sigma', \beta)} m(\gamma) \sum_{\gamma' \in \Gamma_1(\beta, \sigma)} m(\gamma') \\
&= \sum_{\gamma \in \Gamma_r(\sigma', \sigma)} m(\gamma).
\end{aligned}$$
This completes the proof. □

Note that the proof of Theorem 7.20 is technical, but completely elementary. By contrast, the computation of the differential in the smooth Morse complex requires some heavy mathematics, including the construction of a compactification of the space $\mathcal{C}(p,q)$ to a manifold with corners, which we omitted. Thus, if one has a relatively simple discrete Morse function on a complex, it is fairly straightforward to compute homology using the complex $\mathbb{M}_\bullet(X,\mathbb{Z})$.

We conclude this section with an example to illustrate the maps Φ and d and to use the discrete Morse complex to compute the homology of a complex.

Example 7.22. Consider the complex X shown in Figure 7.1. The orientation of each edge is given by the arrows in the figure; these are *not* the arrows showing the gradient of a discrete Morse function. A discrete Morse function is given by the numbers attached to each simplex. Of course, the complex is just a circle, S^1, and we know its homology, but we will compute the discrete Morse complex $\mathbb{M}_\bullet(X,\mathbb{Z})$ and show that it recovers the homology of X.

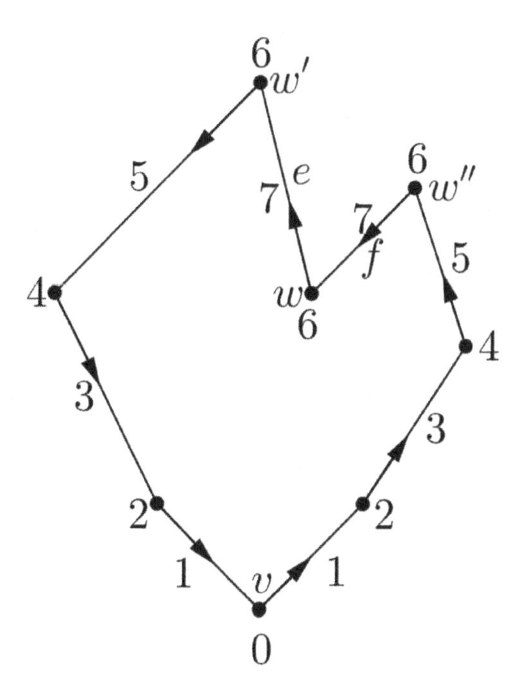

Fig. 7.1 The complex X with a discrete Morse function f

There are two critical vertices, v and w, and two critical edges e and f. The Morse complex $\mathbb{M}_\bullet(X,\mathbb{Z})$ therefore has the form

$$\mathbb{Z}\{e,f\} \xrightarrow{d} \mathbb{Z}\{v,w\},$$

and we may use Theorem 7.20 to compute d. First consider the edge e, whose boundary consists of the vertices w and w': $\partial e = w' - w$. Note that w is a critical vertex and there are no gradient paths from w to v. Note also that there is a single gradient path from w' to v which traverses the left side of the circle from w' down to w. We now calculate:

$$\begin{aligned}
\langle de, v\rangle &= \sum_{u<e} \langle \partial e, u\rangle \sum_{\gamma\in\Gamma_N(u,v)} m(\gamma) \\
&= \langle w' - w, w'\rangle \sum_{\gamma\in\Gamma_N(w',v)} m(\gamma) + \langle w' - w, w\rangle \sum_{\gamma\in\Gamma_N(w,v)} m(\gamma) \\
&= \sum_{\gamma\in\Gamma_N(w',v)} m(\gamma) - \sum_{\gamma\in\Gamma_N(w,v)} m(\gamma) \\
&= (-1)(-1)(-1) - 0 \\
&= -1.
\end{aligned}$$

Similarly, we have the following

$$\begin{aligned}
\langle de, w\rangle &= -1 \\
\langle df, v\rangle &= 1 \\
\langle df, w\rangle &= 1.
\end{aligned}$$

Thus, the matrix of d is

$$\begin{bmatrix} -1 & 1 \\ -1 & 1 \end{bmatrix}.$$

This matrix has rank 1 and its kernel is spanned by $e + f$. The image of d is the span of $v + w$. The calculation of the homology of X follows immediately.

We leave the complete calculation of the invariant complex $K_\bullet(X, \mathbb{Z})$ as an exercise, but we give a partial computation here. Note that Φ fixes all critical cells: $\Phi(v) = v, \Phi(w) = w, \Phi(e) = e, \Phi(f) = f$. Consider the vertex labeled with value 2 on the right half of the circle. Denote this by v_2 and denote by e_1 the edge joining v and v_2. Then

$$\Phi : v_2 \mapsto v_2 + \partial(V(v_2)) = v_2 + \partial(-e_2) = v_2 + (v - v_2) = v.$$

Similarly, denoting by v_4 the vertex labeled with value 4 on the right side and by e_3 the edge joining v_2 and v_4, we have

$$\begin{aligned}
\Phi : e_3 \mapsto e_3 &+ \partial V(e_3) + V(\partial e_3) \\
&= e_3 + 0 + V(v_4 - v_2) \\
&= e_3 + (-e_3 + e_1) \\
&= e_1.
\end{aligned}$$

7.5 Comparison with smooth Morse theory

An obvious question arises: how do these two theories compare? That is, given a smooth manifold M and a Morse function $f : M \to \mathbb{R}$, is there a discrete Morse function on a cell decomposition of M that "mirrors" the behavior of f? The following result gives an answer to this question.

Theorem 7.23. (cf. [Gallais (2010)], Theorem 3.1) *Let M be a smooth manifold and let $F : M \to \mathbb{R}$ be a Morse function with gradient-like vector field v. Then there is a C^1-triangulation of M and a discrete gradient vector field V such that the critical cells of V are in one-to-one correspondence with the critical points of F. Moreover, if σ_p and σ_q are critical cells corresponding to the critical points p and q, respectively, with $\dim \sigma_p = \dim \sigma_q + 1$, then V-paths from the boundary of σ_q to σ_p are in bijection with integral curves of v from q to p.*

There is a small gap in Gallais's proof, but this was corrected by Benedetti [Benedetti (2014)]. Moreover, Benedetti proved the following stronger result.

Theorem 7.24. (cf. [Benedetti (2014)], Theorem 2.28) *Suppose M is a smooth manifold and let $F : M \to \mathbb{R}$ be a Morse function on M with c_i critical points of index i. Then for any PL triangulation T of M, there is an integer $r \geq 0$ such that the rth barycentric subdivision $\operatorname{sd}^r(T)$ supports a discrete Morse function with c_i critical cells of index i.*

Proof. (Sketch) We know that we can decompose M into handles corresponding to the critical points of F. Moreover, each handle contains exactly one critical point in its interior. Recall that a *shelling* of a simplicial complex is an ordering $\sigma_1, \sigma_2, \ldots$ of its maximal simplices such that each complex

$$B_k = \left(\bigcup_{i=1}^{k-1} \sigma_i \right) \cap \sigma_k$$

is pure of dimension $\dim \sigma_k - 1$ ("pure" means all top-dimensional simplices have the same dimension). Not every simplicial complex is shellable, but a result of Adiprasito–Benedetti [Adiprasito and Benedetti (2013)] asserts that for any PL triangulation B of a ball, there is an integer r such that $\operatorname{sd}^r(B)$ is shellable. Shellable balls have the property that if one removes a top dimensional simplex, the resulting subcomplex collapses to the boundary sphere (such complexes are called *endocollapsible*).

Recall that a (discrete) Morse function is *perfect* if $c_i = \beta_i(M)$ for all i. Call a discrete Morse function f on a manifold with boundary M *boundary-critical* if all the cells in ∂M are critical for f. A boundary-critical function f is called *perfect* if the number of critical interior $(d-i)$-cells is $\beta_i(M)$ ($d = \dim M$). A triangulation B of a ball admits a perfect discrete Morse function if and only if B is collapsible, and B admits a perfect boundary-critical discrete Morse function if and only if B is endocollapsible. Thus, given any triangulation of a ball, there is some subdivision that admits a boundary-critical discrete Morse function.

Now, one shows that if N is the union of two shellable balls B_1, B_2 where $B_1 \cap B_2$ is homeomorphic to $S^{i-1} \times I^{d-i}$ for some i, then any boundary-critical discrete Morse function h on $B_1 \cap B_2$ extends to a boundary-critical discrete Morse function $\tilde{h}$ on N whose critical interior cells are exactly those of h minus a $(d-1)$-face, plus a d-face. Inductively, one then begins with a single handle and proves that any d-manifold with boundary admits a boundary-critical discrete Morse function with c_{d-i} interior critical i-faces. Dualizing, we obtain the result for M. □

Remark 7.25. The number r of subdivisions might be large. Even for spheres it is possible to construct PL triangulations for which the sth barycentric subdivision is not endocollapsible.

We will not pursue this comparison further. Instead, we will turn to algorithms for constructing discrete Morse functions and to applications of the theory.

7.6 Exercises

(1) Construct a discrete Morse function on Δ^n with exactly one critical point.
(2) Construct a discrete Morse function on $\partial\Delta^n$ with exactly two critical points.
(3) Prove the following generalization of Theorem 7.2: Suppose V is a discrete vector field on X with Lyapunov function f. Define the discrete sublevel complexes for f just as if f were a discrete Morse function. Then if there are no chain recurrent cells in $f^{-1}([a,b])$, then $X(b) \searrow X(a)$.
(4) Prove the following generalization of Theorem 7.3: Suppose V is a discrete vector field on X with Lyapunov function f. Suppose $f^{-1}([a,b])$ consists of a collection of basic sets. Then $X(b)$ has the homotopy type of $X(a)$ with the basic sets glued in along their boundaries.
(5) Prove the following refinement of Theorem 7.4. Let $f : X \to \mathbb{R}$ be an injective discrete Morse function on X. Let $\sigma_1, \ldots, \sigma_k$ be the critical simplices of f, ordered so that $f(\sigma_1) < \cdots < f(\sigma_k)$. Let $X_i = X(f(\sigma_i))$. Then X collapses to X_k and for each $i = 2, \ldots, k$, σ_i is a maximal simplex in X_i such that $X_i - \sigma_i$ collapses to X_{i-1}. Moreover, $X_1 = \sigma_1$ consists of a single vertex.
(6) Compute the invariant complex $K_\bullet(X, \mathbb{Z})$ for the complex in Example 7.22 and show that it is isomorphic to the discrete Morse complex $\mathbb{M}_\bullet(X, \mathbb{Z})$.
(7) Verify the calculation of the differential in Example 7.22.

Bibliographic notes

Again, much of the material in this chapter comes from Forman's original paper [Forman (1998a)]. Proofs of theorems without explicit reference in the text follow the exposition of that paper.

Chapter 8

Algorithms

Verifying that a given function on the collection of cells of a CW complex is a discrete Morse function requires checking many local combinatorial conditions. In practice, this is time-intensive and tedious, and therefore we wish to find algorithms for constructing these functions. The crucial observation is that one really need only construct a discrete gradient vector field, and this is equivalent to finding an acyclic partial matching on the Hasse diagram of the complex. The latter question is a problem in graph theory, where algorithms abound. In this chapter we will discuss the algorithmic aspects of constructing discrete Morse functions.

A first attempt at constructing a discrete gradient is obvious. Begin with the Hasse diagram of the complex X and start with the empty partial matching $\mathcal{M}$ (Definition 6.37). Order the edges $e_1, \ldots, e_r$, and then proceed as follows: for $i = 1, \ldots, r$, if $\mathcal{M} \cup \{e_i\}$ is acyclic, then add e_i to $\mathcal{M}$; otherwise, leave $\mathcal{M}$ unchanged. This will produce a discrete gradient, but it is likely to contain many critical cells. Moreover, different orderings on the edges will produce different discrete gradients.

What we want in an algorithm is the following:

(1) Near-optimality: a discrete gradient with as few critical cells as possible;
(2) Easy implementation;
(3) Low complexity; e.g., the algorithm runs in polynomial time.

Of course, we are unlikely to get all of these. Indeed, getting optimality and low complexity is impossible in general, as we point out below.

Note: In the remainder of this chapter, X denotes a simplicial complex. Much of what is presented holds for regular CW complexes, but in most applications simplicial complexes are what is used.

8.1 First case: 2-dimensional complexes

Constructing a discrete Morse function f on a 1-dimensional complex is essentially trivial: first compute a spanning tree T; choose an arbitrary root v_0 and set $f(v_0) = 0$; for each vertex v set $f(v)$ to be the distance from v_0 to v; if u and w are adjacent vertices in the tree with $f(w) = f(u) + 1$, and e is the edge joining them, set $f(e) = f(w)$; if e is any edge not in T, set $f(e) = \max\{f(u) : u \in T\} + 1$. One checks easily that f is a discrete Morse function with exactly one critical vertex (v_0) and critical edges exactly the edges not in T. Observe that the function f is optimal in the sense that the number of critical cells equals the Betti number in each dimension.

The first nontrivial case is therefore $\dim X = 2$. The following algorithm is due to Lewiner, et al. [Lewiner et al. (2003a)]. Suppose X is a finite 2-dimensional simplicial complex with the topology of a 2-manifold (possibly with boundary). We first need a definition.

Definition 8.1. Let M be a triangulated n-manifold, possibly with boundary. The *dual pseudograph* of M is the graph whose vertices are the n-simplices and whose edges correspond to the $(n-1)$-simplices adjacent to each n-simplex. Note that since M is a manifold, each $(n-1)$-simplex is a face of either one or two n-simplices; thus the dual pseudograph may have loops (in case an $(n-1)$-simplex is a face of one n-simplex) or multiple edges.

We now proceed as follows.

Algorithm 8.2.

1: Construct a spanning tree T on the dual pseudograph of X.

2: If X has a boundary, add one boundary edge of X to T. This edge will be a loop in the dual pseudograph.

3: Define the discrete Morse function f on T as follows. Choose a root of T and to each vertex in T (which correspond to the 2-simplices in X) assign its distance from the root plus $|X^{(0)}|+1$, the number of vertices in X. Assign to every edge of T (which correspond to some of the 1-cells in X) the minimum value of its two ends.

4: Let G be the complement of T; this is a graph whose nodes are the vertices of X and whose edges are the edges of X not represented in T. Let U be a spanning tree in G and assign to every vertex of G its distance from a chosen root of U. Assign to each edge in G the maximum value of its two ends. For each edge in $G - U$, assign the value $|X^{(0)}|$.

Proposition 8.3. *Algorithm 8.2 generates a discrete Morse function on X.*

Proof. By the construction of f on the trees T and U, it is clear that the inequalities in Definition 6.3 are satisfied. Note that there is exactly one critical vertex, namely the root of U. If $\partial X = \emptyset$, the root of T is the unique critical 2-simplex; otherwise there is no critical 2-simplex in X.

It remains to check the edges. Since we have added the value $|X^{(0)}|$ to the vertices in T, the critical edges are those in $G - U$, which have been assigned a value greater than any of the vertices and less than the value of any 2-simplex. The conditions of Definition 6.3 are clear for the simplices represented in T and U, and every simplex in T has a value greater than any simplex in G. It follows that the conditions are strictly satisfied between the edges of T and the vertices of G and between the edges of U and the 2-simplices in T. □

Algorithm 8.2 is linear in the size of the complex. Once the spanning trees are constructed, each vertex and edge is visited at most once and spanning trees may be constructed in linear time using a simple greedy algorithm. Thus, this is a very efficient algorithm, and we will see in Section 8.4 that it generates an optimal discrete Morse function.

We can extend the algorithm to an arbitrary 2-complex X. Such a complex is not a manifold if any of the following occur.

(1) There is an edge in X not incident to a 2-simplex;
(2) X has a singular vertex; that is, removing the vertex disconnects incident faces;
(3) X has a non-regular edge: an edge incident to 3 or more 2-simplices.

We handle these cases as follows.

(1) If X has an edge that is not part of any 2-simplex, then X is essentially a 2-complex with a graph glued to it. This graph will be processed in the fourth step of Algorithm 8.2 with no difficulty.
(2) If X has a singular vertex, then X is several complexes glued at a vertex. For the first three steps of Algorithm 8.2, each of these complexes is processed as a distinct connected component. Step four generates only one critical vertex.
(3) If X has a non-regular edge, we first remove from the dual pseudograph the edges incident to 3 or more 2-simplices. The algorithm then proceeds as before and the edges that have been removed will be critical.

8.2 General n-complexes

Algorithm 8.2 exploited duality in dimension 2 to build discrete Morse functions. In higher dimensions we must be more creative. In this case we will construct discrete gradient vector fields rather than functions. To do this, recall that a discrete gradient is a partial matching in the Hasse diagram of the complex (Definition 6.37). Observe that in any discrete vector field V, the V-paths are contained entirely in a pair of successive levels of the modified Hasse diagram; that is, a V-path is a directed path in the modified Hasse diagram that alternates between p and $p+1$ cells for some $p = 0, 1, \ldots, \dim X - 1$. Thus, to construct a discrete gradient on X, we need to modify the levels in the Hasse diagram of X in such a way that we do not create directed loops and so that the modified levels are compatible with each other. An efficient algorithm for this is due to Lewiner, et al. [Lewiner et al.

(2003b)].

Definition 8.4. A *hypergraph* is a pair (X, E), where X is a set of elements called vertices, and E is a set of nonempty subsets of X called hyperedges (sometimes called hyperlinks).

Definition 8.5. Hyperedges may be classified into the following categories.

(1) *Regular hyperedges* join two distinct vertices; these correspond to edges in a regular graph.
(2) *Loops* consist of exactly one vertex.
(3) *Nonregular hyperedges* consist of three or more vertices or are multiply incident to one vertex.

Definition 8.6. Let R be the set of regular hyperedges of (X, E). Then the *regular components* of (X, E) are the connected components of the graph (X, R).

Orient the hypergraph (X, E) by labeling one vertex of each hyperedge as its source. This is called a *simple orientation* of the hypergraph.

Definition 8.7. An *oriented hypercircuit* in a hypergraph (X, E) is a sequence of distinct vertices $v_0, v_1, \ldots, v_{r+1}$ such that $v_{r+1} = v_0$ and for all $0 \le i \le r$, v_i is the source of a hyperedge leading to v_{i+1}.

Definition 8.8. A simply oriented hypergraph (X, E) is a *hyperforest* if each vertex is the source of at most one hyperedge and if (X, E) does not contain any oriented hypercircuit.

The proof of the following fact is left as an exercise.

Proposition 8.9. *Let (X, E) be a hyperforest and let R be one of its regular components.*

(1) R is a simple tree.
(2) There is at most one vertex in R which is the source of either a loop or non-regular hyperedge. □

We can represent a hypergraph by a bipartite graph as follows.

Definition 8.10. Let $H = (X, E)$ be a hypergraph. The *associated bipartite graph* $B(H)$ is the bipartite graph whose two classes of vertices are $\{B(v) : v \in X\}$ and $\{B(e) : e \in E\}$. For each $e = \{v_1, \ldots, v_k\} \in E$, there is an edge in $B(H)$ joining $B(e)$ to $B(v_i)$, $i = 1, \ldots, k$. If H is oriented, $B(H)$ is directed as follows.

(1) If a vertex $v \in X$ is the source of a hyperedge e, then $B(e)$ is the source of the edge of $B(H)$ joining $B(e)$ to $B(v)$.
(2) If $v \in X$ is not the source of an incident hyperedge e, then $B(v)$ is the source of the edge of $B(H)$ joining $B(v)$ to $B(e)$.

We can reverse this process.

Definition 8.11. Let $G = (V_1 \coprod V_2, E)$ be a bipartite graph. Then we have two associated hypergraphs: $B^{-1}(G)$ and its dual $D(B^{-1}(G))$. The vertices of $B^{-1}(G)$ are V_1 and the hyperlinks are given as follows: if $v \in V_2$, let $\{e_1, e_2, \ldots, e_k\} \subseteq E$ be the edges incident with v and let $B^{-1}(v) = \{v_1, \ldots, v_k\} \subseteq V_1$ be the set of opposite ends. Then $B^{-1}(v)$ is a hyperlink in $B^{-1}(G)$. The dual $D(B^{-1}(G))$ is obtained by switching the roles of V_1 and V_2.

With these concepts in hand, we may proceed with the development of the algorithm. In the 2-dimensional case, we considered the dual pseudograph of the complex whose vertices are the 2-simplices with edges giving the incidence relations. In general, if X has the topology of an n-manifold, then the top layer of the Hasse diagram, showing the connections between n- and $(n-1)$-simplices is also a pseudograph, which we call the *dual layer* $n/(n-1)$ of the diagram. In general we have the following definition.

Definition 8.12. Let p and q be consecutive dimensions for a complex X (i.e., $|p-q| = 1$). The *layer* $\mathcal{L}_{p/q}$ of the Hasse diagram of X is the bipartite graph whose vertices are the p- and q-cells of K and whose edges join indicent cells.

Remark 8.13. Note that for each pair p, q with $|p-q| = 1$ we have two distinct layers $\mathcal{L}_{p/q}$ and $\mathcal{L}_{q/p}$. They are dual to each other.

Definition 8.14. Let V be a discrete vector field on X and let $\mathcal{L}_{p/q}$ be a layer of the Hasse diagram modified by V. The *reduced layer* $L_{p/q}$ is the oriented bipartite graph obtained from $\mathcal{L}_{p/q}$ by removing

(1) the p-cells of X paired with a q'-cell of X by V, $q' \neq q$;
(2) the unpaired q-cells of X or the q-cells of X paired with a p'-cell by V, $p' \neq p$.

V-paths live in a single reduced layer. Associated to each reduced layer, we have, via Definition 8.11, a hypergraph. If V is a gradient, then this hypergraph will have no hypercircuits.

Definition 8.15. Let V be a discrete vector field on X and let $L_{p/q}$ be a reduced layer of the modified Hasse diagram. The *p/q-hypergraph* of V, $H_{p/q}$ is the hypergraph representation of $L_{p/q}$; that is, $H_{p/q} = B^{-1}(L_{p/q})$. $H_{p/q}$ is oriented by declaring the source of a hyperedge to be the vertex representing its paired cell in V.

Theorem 8.16. *Let V be a discrete vector field on X. Then V is a discrete gradient vector field if and only if the $0/1, 1/2, \ldots, (n-1)/n$ hypergraphs of V are hyperforests.*

Proof. Each vertex in $L_{p/q}$ is the source of at most one hyperedge; we need only show that closed V-paths coincide with hypercircuits in the hypergraphs. This will complete the proof since by Theorem 6.19, V is a gradient if and only if it has no closed V-paths, and this is equivalent to each p/q-hypergraph having no hypercircuits. Suppose $v_0, v_1, \ldots, v_{r+1} = v_0$ is an oriented hypercircuit in $H_{p/q}$. Then v_i is the source of a hyperedge e_i incident to v_{i+1}. The hyperedge e_i represents a q-cell β_i and v_i represents a p-cell α_i. Then α_i, β_i is a pair in V and thus $\alpha_0 < \beta_0 > \alpha_1 < \beta_1 > \cdots < \beta_r > \alpha_{r+1}$ is a V-path. Since $v_{r+1} = v_0$, this is a closed V-path. Reversing this argument we see that a closed V-path is a hypercircuit in one of the $H_{p/q}$. □

Remark 8.17. Since the dual of a hyperforest is a hyperforest, we may replace any element p/q in the sequence $0/1, 1/2, \ldots, (n-1)/n$ by q/p.

Definition 8.18. A regular component of a hyperforest is *critical* if none of its vertices is the source of either a loop or a nonregular hyperedge.

Proposition 8.19. *Let $H_{p/q}$ be the p/q-hyperforest of the discrete gradient V. Then the number of critical components of $H_{p/q}$ equals the number of critical p-cells of V.*

Proof. Each critical p-cell of V corresponds to a vertex in $H_{p/q}$ and its reduced layer $L_{p/q}$. The isolated vertices of $L_{p/q}$ are not paired by V with any other cell in X, and these remain isolated in $H_{p/q}$. These are critical components. By Proposition 8.9, each regular component R is a tree. If such a tree has k vertices, then there are $k-1$ (regular) edges. These are oriented, so there are $k-1$ sources of edges in R and so these cannot be critical. If R is not a critical component there is exactly one vertex of R which is either the source of a loop or a nonregular hyperedge and is therefore not critical. If R is a critical component, then this vertex is not the source of a loop or a nonregular hyperedge. Thus, this vertex is not incident to any regular hyperedge not in R. Since those edges of R are paired with other vertices, this vertex is unpaired in $L_{p/q}$. But then by definition it cannot be paired with a cell outside $L_{p/q}$ and thus it is unpaired; that is, it corresponds to a critical p-cell. □

Algorithm 8.20. Suppose X is an n-dimensional complex and let $H_{p/q}$ be the hypergraph representation of the p/q-layer $\mathcal{L}_{p/q}$ of the Hasse diagram of X. Choose which layers to process; that is, for each p, q with $|p-q| = 1$, choose either p/q or q/p. Construct a hyperforest $F_{p/q}$ in each $H_{p/q}$ as follows.

1: Initialize $F_{p/q}$ to be the vertices of $H_{p/q}$.
2: Generate a spanning tree on each regular component of $H_{p/q}$.
3: Add the hyperedges of each spanning tree to $F_{p/q}$.
4: If a regular component of $H_{p/q}$ is incident to some loops, add one of them to $F_{p/q}$.
5: Add the nonregular hyperedges of $H_{p/q}$ which do not create cycles.

Note that the first four steps of the algorithm are linear in the size of the complex X. Only the last step creates complexity. Indeed, there is no polynomial algorithm to achieve the fifth step. If the hypergraph is small enough, an exponential time algorithm may be used: generate all possible hyperforests and search for a maximal one. In Section 8.4 we discuss optimality for Algorithm 8.20.

Theorem 8.16 tells us that a hyperforest in each layer p/q yields a discrete gradient vector field. Thus, the collection of $F_{p/q}$ gives us a gradient on X. Note also that Algorithm 8.2 is contained in Algorithm 8.20, at least implicitly. Indeed, the former algorithm is based on finding spanning trees in $\mathcal{L}_{0/1}$ and $\mathcal{L}_{2/1}$ and then piecing them together into a discrete Morse function.

8.3 From point data to discrete Morse functions

In this section we address the following question. Say we have a simplicial complex with a function defined on the set of vertices. The most obvious situation is in data analysis: we have a discrete sample of function values on some space, which we could then triangulate with the sample points as vertices. Can we extend this function to a discrete Morse function on the entire complex in a reasonable way? That is, if we are sampling a smooth function from some manifold, for example, can we extend this to a discrete Morse function which mirrors the behavior of the smooth one?

A trivial extension certainly exists: let M be the maximum value of the function on the vertices and then assign the value $M + \dim \sigma$ to each simplex σ. This makes every cell critical, so this is not what we want. We now describe an algorithm, due to King, et al. [King et al. (2005)], which generates a discrete gradient from the point data with relatively few critical cells.

Let K be a finite simplicial complex and denote by K_i the set of i-simplices of K. Suppose $h : K_0 \to \mathbb{R}$ is an injective function (in practice, if we sample a function on a space we may not have this condition, but it is possible to perturb the function slightly to make it injective on the sample). If $\sigma = \langle v_0, v_1, \ldots, v_i \rangle$ is an i-simplex, define

$$\mathrm{maxh}(\sigma) = \max_{0 \le j \le i} \{h(v_j)\}.$$

Recall that if $\sigma \in K_i$ and $\tau \in K_j$ are disjoint simplices, the *join* $\sigma * \tau$ is either undefined or the $(i + j + 1)$-simplex whose vertices are the union of those of σ and τ (see Appendix B).

Definition 8.21. Let $v \in K_0$ be a vertex of K. The *link* of v is the simplicial complex L whose simplices are all $\tau \in K$ such that $v * \tau$ is defined. The *lower link* of v is the maximal subcomplex of the link of v whose vertices have h-value less than $h(v)$. That is, the simplices in the lower link are all $\tau \in K$ such that $v * \tau$ is defined and $\mathrm{maxh}(\tau) < h(v)$.

See Figure 8.1 for an example of the lower link of a vertex. The values of h on the vertices around v are indicated $\pm$ depending on whether the value is greater than or less than that of $h(v)$.

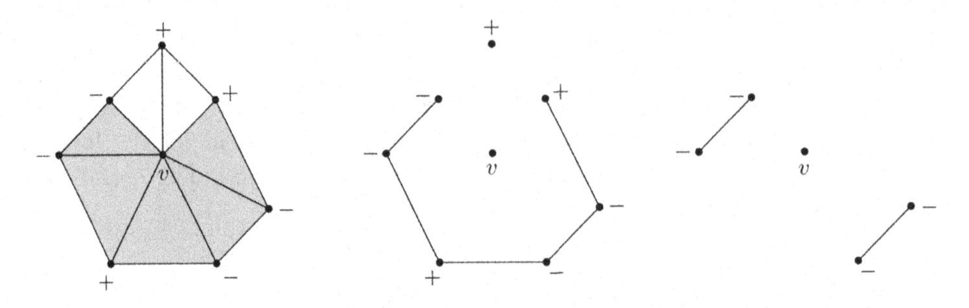

Fig. 8.1 A vertex v in K (left), its link (center), and lower link (right)

We are now ready to describe the algorithm $\texttt{Extract}(K, h, p)$ with input a finite simplicial complex K, an injective function $h : K_0 \to \mathbb{R}$, and a parameter $p \geq 0$ called *persistence*. The output is

(1) Three sets A, B, C which partition the simplices of K;
(2) A bijection $r : B \to A$ such that $r(\sigma)$ is a codimension-one face of σ.

This information encodes a discrete vector field on K–the simplices in C are critical and the map r gives the pairing between regular simplices. We will use this information to extend the function h to a discrete Morse function on K.

We will need to consider subgraphs R_i of the modified Hasse diagram of K, $i = 1, \ldots, \dim K$. Directed paths in R_i connecting critical i and $i-1$ simplices correspond to gradient paths between those simplices (we called the R_i the layers of the Hasse diagram in Section 8.2). For each i, let $A_i = A \cap K_i$, $B_i = B \cap K_i$, and $C_i = C \cap K_i$. The vertices in R_i are of two types: the $(i-1)$-simplices in $A_{i-1} \cup C_{i-1}$ and the i-simplices in $C_i \cup B_i$. Edges are directed from a simplex to its face except for $\sigma \in B_i$ in which case the edge between $r(\sigma)$ and σ is directed from $r(\sigma)$ to σ. Note that there are two types of initial vertices in R_i: i-simplices in C_i and $(i-1)$-simplices $\sigma \in A_{i-1}$ whose only coface in $C_i \cup B_i$ is $r^{-1}(\sigma)$. The terminal vertices in R_i also come in two types: $(i-1)$-simplices in C_{i-1} and i-simplices τ all of whose $(i-1)$-faces are in B_{i-1} except for the face $r(\tau)$.

The algorithm $\texttt{Extract}$ consists of two parts. $\texttt{ExtractRaw}(K, h)$ generates a preliminary partition A, B, C and function r. Then, the routine $\texttt{ExtractCancel}(K, h, p, j)$ cancels pairs of critical j and $(j-1)$-simplices whose maxh-values differ by at most p. The pseudocode for the algorithms is as follows.

Algorithm 8.22. $\texttt{Extract}(K, h, p)$

1: $\texttt{ExtractRaw}(K, h)$
2: **for** $j = 1 = 1, \ldots, \dim K$ **do**
3: $\quad\texttt{ExtractCancel}(K, h, p, j)$
4: **end for**

Algorithm 8.23. `ExtractRaw`(K, h)

1: Initialize A, B, C to be empty.
2: **for all** $v \in K_0$ **do**
3: let K' = the lower link of v
4: **if** K' is empty **then** add v to C ▷ local min
5: **else**
6: Add v to A.
7: Let $h' \colon K'_0 \to \mathbb{R}$ be the restriction of h, or alternatively use the definition given in the text below.
8: `Extract`(K', h', ∞) and let A', B', C', r' denote the resulting partition of the simplices of K'
9: find the $w_0 \in C'_0$ so that $h'(w_0)$ is the smallest. Add $[v, w_0]$ to B and define $r([v, w_0]) = v$.
10: for each $\sigma \in C' - w_0$ add $v * \sigma$ to C.
11: for each $\sigma \in B'$ add $v * \sigma$ to B, add $v * r'(\sigma)$ to A and define $r(v * \sigma) = v * r'(\sigma)$.
12: **end if**
13: **end for**

If we have a metric on K which gives each edge a length then we have the following alternative definition of h' in the lower link of v with the property that the vertex with the minimum value of h' more closely approximates the direction of steepest decrease of h: $h'(w) = (h(w) - h(v))/\ell([v, w])$, where $\ell([v, w])$ is the length of the edge $[v, w]$. In case this produces a non-injective h', perturb h' slightly, breaking ties using the value $h(w)$.

`ExtractRaw` may produce many critical cells. The routine `ExtractCancel` removes those cells that are connected by a single gradient path and whose values differ by a specified parameter.

Algorithm 8.24. `ExtractCancel`(K, h, p, j)

1: **for all** $\sigma \in C_j$ **do**
2: Find all gradient paths $\sigma = \sigma_{i1} \to \sigma_{i2} \to \ldots \to \sigma_{i\ell_i} \in C_{j-1}$ with $\operatorname{maxh}(\sigma_{i\ell_i}) > \operatorname{maxh}(\sigma) - p$
3: **for all** i **do** if $\sigma_{i\ell_i}$ does not equal any other $\sigma_{j\ell_j}$ let $m_i = \operatorname{maxh}(\sigma_{i\ell_i})$
4: **if** at least one m_i is defined **then**
5: choose j with $m_j = \min\{m_i\}$
6: `Cancel`$(K, h, \sigma_{j\ell_j}, \sigma, j)$
7: **end if**
8: **end for**
9: **end for**

Finally, we describe the algorithm `Cancel`(K, h, σ, τ, j), which works if $\sigma \in C_{j-1}$, $\tau \in C_j$ and there is exactly one gradient path from τ to σ. It reverses the arrows in the gradient path, thereby removing the two critical cells.

Algorithm 8.25. `Cancel`(K, h, σ, τ, j)

1: Find the unique gradient path $\tau = \tau_1 \to \sigma_1 \to \tau_2 \to \sigma_2 \to \cdots \to \sigma_k = \sigma$
2: Delete σ and τ from C, add τ to B, and add σ to A
3: **for** $i = 1, \ldots k$ **do** redefine $r(\tau_i) = \sigma_i$
4: **end for**

Theorem 8.26. *The A, B, C and r produced by* `ExtractRaw` *have the property that there are no directed loops in the resulting modified Hasse diagram.*

Proof. Note that $\text{maxh}(r(\sigma)) = \text{maxh}(\sigma)$ since in the algorithm v is the vertex in all mentioned simplices with highest value of h. Also if σ is a face of τ then $\text{maxh}(\sigma) \leq \text{maxh}(\tau)$. Consequently maxh is nonincreasing along any directed path in the modified Hasse diagram. So maxh must be constant on any directed loop. Let $\sigma_0 \to \sigma_1 \to \cdots \to \sigma_k = \sigma_0$ be a directed loop in the modified Hasse diagram. Let v be the unique vertex so $h(v) = \text{maxh}(\sigma_j)$ for all j. Then v is a vertex of each σ_j and hence $\sigma_j = v * \tau_j$ for simplices τ_j in the lower link of v. But then $\tau_0 \to \tau_1 \to \cdots \to \tau_k$ is a directed loop in the modified Hasse diagram of the lower link of v, which is impossible by induction on dimension and the following theorem. □

Theorem 8.27. *The* `Cancel` *algorithm does not produce directed loops; thus the A, B, C and r produced by* `Extract` *have the property that there are no directed loops in the resulting modified Hasse diagram.*

Proof. (cf., Theorem 7.12) Suppose we have A, B, C and r so the resulting modified Hasse diagram has no directed loops. Suppose $\tau, \sigma \in C$ are joined by a unique gradient path. Assume that after performing `Cancel`(K, h, σ, τ, j) we have A', B', C' and r' such that the resulting modified Hasse diagram has a directed loop α. This was not previously a directed loop and hence a portion of it (say γ) must coincide with a segment (say η) of the gradient path. But then we can construct a different gradient path from τ to σ by replacing η by $\alpha - \gamma$. This violates the condition that there be only one gradient path from τ to σ. □

Our goal was to produce a discrete Morse function that mirrors the behavior of our function h on the vertices. The algorithms above generate a discrete gradient vector field, and we now show how to extend h to a discrete Morse function having this vector field as its gradient.

Theorem 8.28. *There is an extension of h to a discrete Morse function h' with the same modified Hasse diagram as that produced by* `ExtractRaw`*. Moreover, given $\varepsilon > 0$ we may choose such an h' so that $|h'(\tau) - \text{maxh}(\tau)| \leq \epsilon$ for any simplex τ.*

Proof. We may as well suppose that $3\epsilon < |h(v) - h(w)|$ for all vertices $v \neq w$. If K' is the lower link of a vertex v, we may by Theorem 8.26 find a discrete Morse function g_v on K'. After a linear scaling we may suppose the range of g_v is in the interval $(h(v), h(v)+\epsilon]$. Let w_0 be the vertex of K' which minimizes h. By Lemma 8.29 below we know that w_0 is critical in K'. Now define h' on the lower star of v by $h'([v, w_0]) = h(v) - \epsilon$ and $h'(v * \tau) = g_v(\tau)$ for any simplex $\tau \neq w_0$ of K'.

We claim that if τ is a codimension one face of σ and $h'(\tau) \geq h'(\sigma)$ then $\sigma \in B$ and $\tau = r(\sigma)$. Conversely, we claim that $h'(r(\sigma)) \geq h'(\sigma)$ for all $\sigma \in B$. The first claim implies that h' is a discrete Morse function, and the second then implies that the modified Hasse diagram for this Morse function coincides with that produced by `ExtractRaw`.

Let us prove the claims. Let v be the maximal vertex of σ so $h(v) = \text{maxh}(\sigma)$. If $\sigma \neq v$ let σ' be the simplex so $\sigma = v * \sigma'$. To prove the first claim, let w be the maximal vertex of τ. If $w \neq v$ then $h'(\tau) \geq h'(\sigma) \geq h(v) - \epsilon \geq h(w) + 2\epsilon \geq h'(\tau) + \epsilon$, a contradiction, so in fact $w = v$, i.e., τ is in the lower star of v. Suppose first that $\tau = v$. Then the only possibility is $\sigma = [v, w_0]$ so $r(\sigma) = \tau$. Now suppose that $\tau \neq v$. Then there is a simplex τ' in the lower link of v so that $\tau = v * \tau'$. So $g_v(\tau') \geq g_v(\sigma')$ and thus there is only one possibility for τ' and in fact $\tau = r(\sigma)$. Now let us prove the second claim. It holds if $r(\sigma) = v$ since then $\sigma = [v, w_0]$ and $h'(\sigma) = h(v) - \epsilon$. But if $r(\sigma) \neq v$ then $r(\sigma) = v * r'(\sigma')$ so $h'(r(\sigma)) = g_v(r'(\sigma')) \geq g_v(\sigma') = h'(\sigma)$. □

A similar result does not hold for `Extract`. Recall that in the smooth case canceling pairs of critical points requires changing the values of the Morse function; the same is true for discrete Morse functions.

Lemma 8.29. *If v is the vertex of K at which h attains its minimum then* `Extract` *will make v a critical vertex.*

Proof. `ExtractRaw` will make v critical because it is a local minimum. But then `ExtractCancel` will never cancel v: any critical 1-simplex τ is the start of exactly two gradient paths, so if τ is connected to v by a single gradient path it must be connected to some other vertex w by a single gradient path. Since $\text{maxh}(\tau) - h(w) < \text{maxh}(\tau) - h(v)$, `ExtractCancel` would cancel τ with w instead of with v, if it cancelled at all. □

8.4 Optimality

Recall that a (discrete) Morse function is *perfect* if the number m_i of critical i-cells equals the ith Betti number β_i. From the point of view of computing homology, this is exactly what we want since the differential in the resulting Morse complex vanishes. Finding perfect Morse functions is difficult in general; in this section we discuss how close the algorithms of the previous sections come to achieving optimality.

Recall that a decision problem is in NP if any proposed solution can be verified with a polynomial time algorithm. A problem is *NP-hard* if there is a polynomial time reduction algorithm to any problem in NP. Examples of NP-hard problems include the traveling salesman problem and the graph isomorphism problem. Also, any NP-complete problem (i.e., those problems C which are in NP and such that any problem in NP is reducible to C in polynomial time) is NP-hard. Such a problem is *strongly NP-complete* if it is still NP-complete even when its input parameters are bounded by a polynomial in the length of the input.

Eğecioğlu and Gonzalez [Eğecioğlu and Gonzalez (1996)] proved the following.

Theorem 8.30. *Let X be a connected pure 2-dimensional simplicial complex embeddable in $\mathbb{R}^3$ and let k be a nonnegative integer. Then it is strongly NP-complete to decide whether there exists a discrete gradient vector field on X with at most k critical 2-simplices.* □

Remark 8.31. Actually, in [Eğecioğlu and Gonzalez (1996)] the authors prove the strong NP-completeness of a collapsibility problem: given a set Y of 2-simplices in X with $|Y| \leq k$, is there a sequence of collapses from $X - Y$ to a 1-dimensional complex? Theorem 8.30 is the discrete Morse theoretic reformulation of this result.

We now show that computing an optimal discrete gradient on a complex is NP-hard, using an argument of Joswig and Pfetsch [Joswig and Pfetsch (2006)]. First, a lemma.

Lemma 8.32. *Let X be a connected simplicial complex of dimension at least 1 and let M be a Morse matching on X. Let $\Gamma(M)$ be the graph obtained from the 1-skeleton of X by removing all the edges matched with 2-simplices by M. Then $\Gamma(M)$ is connected.*

Proof. We may assume that $\dim X \geq 2$. Suppose that $\Gamma(M)$ is not connected and let N be the set of vertices in a connected component. Let C be the set of cut edges (those edges of X with one vertex in N and one vertex in the complement of N). Since X is connected, C is nonempty. By the definition of $\Gamma(N)$ each edge in C is matched to a unique 2-simplex. Let D be the directed subgraph of the modified Hasse diagram of X consisting of the edges in C and their matching 2-simplices. Consider the following path in D: start with any node of D corresponding to a cut edge $e_1 \in C$. Let τ_1 be the unique 2-simplex paired with e_1 by M. Then τ_1 contains at least one other cut edge e_2 (otherwise e_1 would not be a cut edge). Pass to e_2 and then its unique paired co-face τ_2. This gives a directed path in D.

Since the graph is finite, this path must eventually return to some node. But then D (and also the modified Hasse diagram) has a directed cycle, contrary to the fact that M is a Morse matching. □

Corollary 8.33. *Let M be a Morse matching on X. Then in polynomial time we can modify M to another Morse matching M' with exactly one critical vertex, the same number of critical faces of dimension 2 or higher, and such that $m_1(M') \leq m_1(M)$.*

Proof. Choose an arbitrary vertex v and any spanning tree of $\Gamma(M)$ (which may be computed in polynomial time). Direct all edges away from v; this gives a maximal Morse matching on $\Gamma(M)$ with a single critical vertex. Replacing M on $\Gamma(M)$ with this matching still yields a Morse matching on X. Note that the total number of critical edges can only decrease in this way since we computed an optimal Morse matching on $\Gamma(M)$. The number of critical i-cells for $i \geq 2$ remains unchanged. □

Theorem 8.34. *Let X be a simplicial complex and let c be a nonnegative integer. Then it is strongly NP-complete to decide whether there is a Morse matching with at most c critical cells, even if X is a pure 2-dimensional complex embeddable in $\mathbb{R}^3$.*

Proof. This is clearly in NP. Let (X, k) be an input for the collapsibility problem (Remark 8.31). We will show that there is a Morse matching with at most k critical 2-simplices if and only if there is a Morse matching with at most $g(k) = 2(k+1) - \chi(X)$ critical cells. Since the Euler characteristic may be computed in polynomial time, g is a polynomial-time computable function. Theorem 8.30 then implies the result.

Assume M is a Morse matching on X with at most k critical 2-simplices. Then by Corollary 8.33 we obtain a Morse matching M' (in polynomial time) with $m_0(M') = 1$, $m_2(M') = m_2(M)$, and $m_1(M') \leq m_1(M)$. By the Morse inequalities, $c_1(M') = c_2(M') + 1 - \chi(X)$. Let $c(M') = c_0(M') + c_1(M') + c_2(M')$. Then

$$c_2(M) = c_2(M') = \frac{1}{2}\Big(c(M') + \chi(X)\Big) - 1.$$

It follows that $c(M') \leq 2(k+1) - \chi(X)$.

Conversely, if M is a Morse matching with at most $g(k)$ critical cells, then computing M' as above we find that

$$c_2(M) = c_2(M') \leq \frac{1}{2}\Big(g(k) + \chi(X)\Big) - 1 = k,$$

and this finishes the proof. □

In practice, however, the algorithms of the preceding sections run quickly and often yield optimal matchings. We have the following results.

Proposition 8.35. *Algorithm 8.2 yields an optimal discrete Morse function on the 2-manifold X.*

Proof. Note that the discrete Morse function f generated by Algorithm 8.2 has exactly 1 critical vertex and 1 critical 2-simplex only if X has no boundary (if $\partial X \neq \emptyset$ there are no critical 2-simplices). But since $\chi(X)$ is the alternating sum of both the Betti numbers and numbers of critical simplices, we must have exactly $\beta_1(X)$ critical edges. □

As for Algorithm 8.20 we have the following.

Proposition 8.36. *The first four steps of Algorithm 8.20 run in linear time and are optimal. The last step, deciding which nonregular hyperlinks to add, need not be optimal.*

Proof. That the first four steps are optimal is clear since they involve finding spanning trees and adding a few loops. The problem with the last step is that at each stage we must make sure a particular hyperlink will not create circuits. There is no polynomial algorithm to add the maximal number of such hyperlinks and any polynomial approximation may terminate before achieving optimality. □

Algorithm 8.22 can be very slow in worst cases since it works inductively on the lower links of vertices in the complex. In practice, however, it yields results quickly with relatively few critical cells. An implementation for subcomplexes of $\mathbb{R}^3$ is available at `http://www-users.math.umd.edu/~hking/MorseExtract.html`.

Applying `Extract` (Algorithm 8.22) with $p = \infty$ in low dimensions yields the minimum possible number of critical simplices. In particular, we have the following result.

Proposition 8.37. *Suppose K has dimension less than 2, or K has dimension 2 and is a subcomplex of a 2-dimensional manifold. If one applies* `Extract` *with $p = \infty$, then the number of critical i-simplices is the rank of $H_i(|K|;\mathbb{Z}/2)$ and hence must be minimal.*

Proof. By Lemma 8.39 below, this result is true for $i = 0$ and 2. But the Euler characteristic $\chi(|K|)$ is both the alternating sum of the ranks of $H_i(|K|;\mathbb{Z}/2)$ and so as in Proposition 8.35 the result must be true for $i = 1$. □

Corollary 8.38. *Suppose $|K|$ is a manifold of dimension $n \leq 3$ and suppose for every vertex v either*

(1) the lower link of v is empty (local min), or
(2) the lower link of v deformation retracts to a $(k-1)$-sphere (local max if $k = n$, otherwise an index k saddle), or
(3) the lower link of v deformation retracts to a point (regular point).

Then when we apply `ExtractRaw` *we will obtain exactly one critical simplex for each vertex which is not a regular point. Each local minimum will be a critical vertex. Each local maximum will be in a critical n-simplex. Each index k saddle will be in a critical k-simplex.*

Proof. If the lower link of v is empty, `ExtractRaw` will designate v a critical 0-simplex. If the lower link of v deformation retracts to a point then by Theorem 8.37 when we apply `Extract` to the lower link of v we will get just one critical simplex, a critical vertex, which does not produce any critical simplices of K. If the lower link of v deformation retracts to a $(k-1)$-sphere then when we apply `Extract` to the lower link of v we will get just two critical simplices, one a critical vertex and the other a critical $(k-1)$-simplex, which produce a critical k-simplex of K. □

Lemma 8.39. *Applying* `Extract` *to* K *with* $p = \infty$ *produces exactly one critical* 0*-simplex in each connected component of* $|K|$*. If* K *is a subcomplex of an* n*-dimensional manifold, then there will be exactly one critical* n*-simplex in each component of* $|K|$ *which is itself an* n*-manifold without boundary. In particular, the number of critical* n*-simplices is the rank of* $H_n(|K|;\mathbb{Z}/2)$.

Proof. Each critical 1-simplex is the origin of only two gradient paths. These paths could not end up at different critical 0-simplices, since `Extract` would have canceled one of them. So both paths end up at the same critical 0-simplex, which implies that $\partial_1 = 0$ in the discrete Morse complex (Theorem 7.20). It follows that $H_0(|K|;\mathbb{Z}/2) = \mathbb{M}_0$ has rank equal to the number of critical 0-simplices; i.e., there is exactly one critical 0-simplex in each component of $|K|$.

Now suppose K is a subcomplex of an n-dimensional manifold. Then each $(n-1)$-simplex is a face of at most two n-simplices. Consequently, each critical $(n-1)$-simplex is the end of at most two gradient paths. So again, $\partial_n = 0$ and $H_n(|K|;\mathbb{Z}/2) = \mathbb{M}_n$. □

8.5 Exercises

(1) Consider the triangulation of the torus T shown below. Use Algorithm 8.2 to construct a discrete Morse function on T.

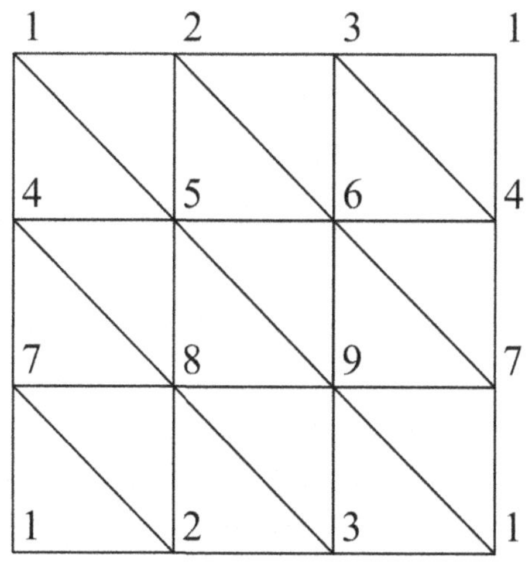

(2) Consider the triangulation of the real projective plane $\mathbb{R}P^2$ shown below. Use Algorithm 8.2 to construct a discrete Morse function on $\mathbb{R}P^2$.

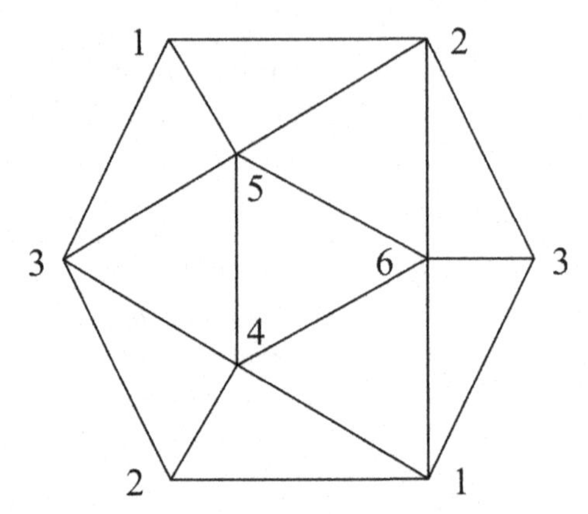

(3) Remove the center triangle from $\mathbb{R}P^2$ and use Algorithm 8.2 to construct a discrete Morse function on the resulting complex.

(4) Prove Proposition 8.9.

(5) Consider the torus in Figure 6.2. Beginning with the values of the function on the vertices, use Algorithm 8.22 to generate a discrete Morse function on the torus. Do you get the same critical simplices?

Chapter 9

Applications

Now that we have the tools of discrete Morse theory in hand we present several examples.

9.1 Combinatorial applications

Given the combinatorial nature of the definition of a discrete Morse function, it should be unsurprising that there are many applications in the areas of graph theory and combinatorics. We begin with the complex of not-connected graphs.

Example 9.1. The complex of not-connected graphs. Consider the complete graph K_n on n vertices. A *spanning subgraph* is a subgraph $G \subseteq K_n$ containing all n vertices. Let N_n be the collection of all spanning subgraphs which are not connected. Observe that if $G_1 \subset G_2$ in K_n and $G_2 \in N_n$, then $G_1 \in N_n$; that is, a spanning subgraph of a not connected graph is also not connected.

Using N_n, we construct a simplicial complex X_n as follows. The k-simplices of X_n are the graphs in N_n which have $k+1$ edges. If G is such a graph, then the faces of G are all the nontrivial spanning subgraphs of G. The structure of the complex X_n is easily described and we shall use discrete Morse theory for the proof. The proof presented below is from Forman's paper [Forman (2002b)].

Theorem 9.2. *The space X_n has the homotopy type of a wedge of $(n-1)!$ copies of the sphere S^{n-3}.*

Proof. Note that the statement of the theorem implies that we need only consider $n \geq 3$. Indeed, if $n = 1$, then the set $N_1 = \emptyset$. If $n = 2$, the set N_2 consists of a single element, namely the graph consisting of two vertices and no edges. The complex X_2 is therefore a single point.

Assume $n \geq 3$ and construct a discrete vector field V on X_n iteratively as follows. Label the vertices of any graph G as $\{1, 2, \ldots, n\}$ and denote by e_{ij} the edge joining vertices i and j (note that a particular graph may or may not contain e_{ij}). Let V_{12} be the vector field consisting of all pairs $\{G < G + e_{12}\}$ for graphs

$G \in N_n$ with $e_{12} \notin G$ and $G + e_{12} \in N_n$. An alternate description is that if $e_{12} \in G$, then V_{12} pairs $G - e_{12}$ with G. Note that the graph $G_0 = e_{12}$ must remain unpaired since the graph $G_0 - e_{12}$ is the empty graph, corresponding to the empty simplex in X_n. The graphs in X_n other than G_0 which are not paired in V_{12} are those G not containing e_{12} with $G + e_{12}$ connected. Note that such a G has exactly two connected components, one containing vertex 1 and one containing vertex 2; denote the components by G_1 and G_2.

Consider such a graph G. Note that vertex 3 must lie in either G_1 or G_2. If it lies in G_1 and G does not contain e_{13} then $G + e_{13}$ is unpaired in V_{12}; add the pair $\{G < G + e_{13}\}$. If G contains e_{13} then G remains unpaired if and only if $G - e_{13}$ is the union of three connected components, each containing one of vertices 1, 2, or 3. If vertex 3 is in G_2 and $e_{23} \notin G$, then add the pair $\{G < G + e_{23}\}$. Denote the resulting vector field by V_3. The unpaired graphs in V_3 are G_0 and those G that either (a) contain e_{13} and are such that $G - e_{13}$ is the union of three connected components, each containing one of the vertices 1, 2, or 3, or (b) contain e_{23} and are such that $G - e_{23}$ is the union of three connected components, each containing one of the vertices 1, 2, or 3.

Now consider vertex 4 and pair any G which is unpaired in V_3 with $G + e_{14}$, $G + e_{24}$, or $G + e_{34}$ if possible (at most one of these graphs is unpaired in V_3). This gives the vector field V_4. Continue in this manner to generate $V = V_n$. The only unpaired graphs in V are G_0 and those graphs that are the union of two connected trees, one containing vertex 1 and one containing vertex 2. There are $(n-1)!$ such graphs, each having $(n-2)$ edges; they therefore correspond to $(n-3)$-simplices in X_n.

We claim that V is a discrete gradient. Note that V_{12} has no closed paths: if $\gamma = \alpha_0^{(p)} < \beta_0^{(p+1)} > \alpha_1^{(p)}$ is a V_{12}-path, then $\beta_0 = \alpha_0 + e_{12}$. Since α_1 is a face of β_1, we must have $\alpha_1 = \alpha_0 + e_{12} - e$ for some edge $e \neq e_{12}$. Since $e_{12} \in \alpha_1$, it is paired with a smaller graph by V_{12} and so γ cannot be continued to a longer V_{12}-path. This implies that there are no closed V_{12}-paths. In general, suppose we have a V-path $\gamma = \alpha_0 < \beta_0 > \alpha_1$ and say α_0 and β_0 are first paired in V_i, $i \geq 3$. Then either α_1 is paired with a smaller graph by V_i, in which case γ cannot be continued, or α_1 is already paired in V_{i-1}. Since V_{i-1} is a gradient by the inductive hypothesis, we see that V has no closed paths.

The result now follows by applying Theorem 7.4. □

One may use similar techniques to study the structure of the complex of not i-connected graphs for $i > 1$, see [Vassiliev (1993)], [Jonsson (2008)], [Shareshian (2001)].

Example 9.3. Discrete Morse functions from lexicographic orderings. In Chapter 8 we saw several algorithms for constructing discrete Morse functions on simplicial complexes. Here we consider a result of Babson and Hersh [Babson and Hersh (2005)] which constructs a discrete Morse function on the order complex,

$\Delta(P)$, of a partially ordered set P with unique minimal and maximal elements, $\hat{0}$ and $\hat{1}$, via lexicographic orderings on the maximal simplices (facets) of $\Delta(P)$. Recall that $\Delta(P)$ contains an i-simplex for each chain $\hat{0} < v_0 < \cdots < v_i < \hat{1}$ in P. This may seem rather specialized, but it may be used for arbitrary simplicial complexes by noting that the order complex of the face poset of a simplicial complex X is the first barycentric subdivision of X. This construction is therefore useful in great generality.

We begin with some definitions.

Definition 9.4. A chain $\hat{0} = v_0 < v_1 < \cdots < v_r = \hat{1}$ is *saturated* (or *maximal*) if $v_i \leq v \leq v_{i+1}$ implies $v = v_i$ or $v = v_{i+1}$ for $0 \leq i < r$. A poset P is *graded* if for each $x \in P$, all saturated chains from $\hat{0}$ to x have the same length, called the *rank* of x.

Definition 9.5. A poset *lexicographic order* is a total ordering of the facets of $\Delta(P)$ with the following property. Suppose two facets F_1, F_2 share a face σ of ranks $1, 2, \ldots, i$, and let $\tau \subseteq F_1$ and $\mu \subseteq F_2$ be faces of ranks $1, 2, \ldots, i+1$ with $\tau \neq \mu$. If F_1 precedes F_2, then any facet containing τ must come before any facet containing μ.

Remark 9.6. One may construct such an order as follows. The Hasse diagram of P is the graph whose vertices are elements of P and whose edges are pairs (u, w) of comparable poset elements with no intermediate comparable elements. Denote such a minimal comparability by $u \prec w$ and call it a *covering relation.* Label the edges in the Hasse diagram with positive integers and then lexicographically order the sequences of labels associated to the saturated chains. Such a labeling is called an *edge-labeling.*

We now describe the matching associated to a lexicographic ordering $F_1, \ldots, F_k$. Denote this order by $<_{\text{lex}}$. If σ is a simplex in $\Delta(P)$, denote by $e\sigma$ the lexicographically earliest facet containing σ. If $e\sigma <_{\text{lex}} e\tau$, then we write $\sigma <_{\text{lex}} \tau$. Note that each maximal face in $F_j \cap (\cup_{i<j} F_i)$ consists of a subchain of F_j given by skipping a single interval of consecutive ranks in F_j. Call such intervals in ranks *minimal skipped intervals* of F_j and denote this set of intervals by $I(F_j)$. The number of ranks which are skipped to obtain such a maximal face is called the *height* of the interval. A minimal skipped interval is *nontrivial* if it skips more than one rank.

A face in F_j belongs to a lexicographically earlier facet if it is disjoint from all of the minimal skipped intervals of F_j. It follows that the faces in $F_j - (\cup_{i<j} F_i)$ are those faces in F_j which hit every minimal skipped interval of F_j. We now alter the set $I(F_j)$ to produce a collection $J(F_j)$ of disjoint intervals and use these to define a matching M on $\Delta(P)$.

Algorithm 9.7.

1: Initialize $J = \emptyset$.

2: Add to J the interval $(u, v) \in I$ with u of smallest possible rank.
3: Replace I by the restriction of I to ranks above the rank of v.
4: Delete from I the skipped intervals which are no longer minimal.
5: Repeat until $I = \emptyset$.

If I does not include all the vertices of F_j, let j_0 be the set of uncovered elements. If $j_0 \neq \emptyset$, let ρ_0 denote the lowest rank element whose rank is in j_0. For each face $\sigma \in e^{-1}(F_j)$ containing the element ρ_0, add the pair $\{\sigma - \{\rho_0\} < \sigma\}$ to M. For $i > 0$, let ρ_i be the lowest rank element from the interval $j_i \in J$ and let $[r]$ be the set of indices on the intervals in J. Define a map

$$t : e^{-1}(F_j) \to [r] \cup \{\infty\}$$

by

$$t(\sigma) = \min_i\{\sigma \cap j_i \neq \{\rho_i\}\},$$

setting $t(\sigma) = \infty$ when the set $\{\sigma \cap j_i \neq \{\rho_i\}\}$ is empty. If $\rho_{t(\sigma)} \in \sigma$ then add $\{\sigma - \{\rho_{t(\sigma)}\} < \sigma\}$ to M; if $\rho_{t(\sigma)} \notin \sigma$ then add $\{\sigma < \sigma \cup \{\rho_{t(\sigma)}\}\}$ to M. Note that if $t(\sigma) = \infty$ then σ is unmatched; this happens when $j_0 = \emptyset$ and σ consists of exactly the minimal elements ρ_i of each interval.

Before proving that this matching M is acyclic we pause to give an example to illustrate the algorithm. Let P be the face poset of the boundary X of the standard 2-simplex. The complex $\Delta(P)$ is then the first barycentric subdivision of X. We give an edge-labeling of P in Figure 9.1.

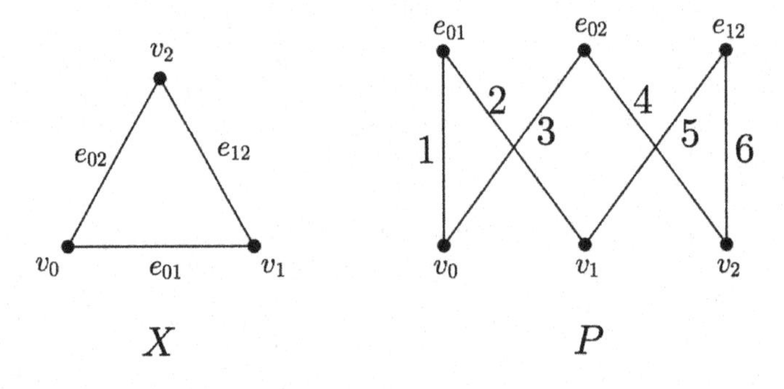

Fig. 9.1 The space X (left) and its face poset P (right)

In this case, we have the following sets $I(F_j)$:

$$I(F_1) = \emptyset$$
$$I(F_2) = \{e_{01}\}$$
$$I(F_3) = \{v_0\}$$
$$I(F_4) = \{e_{02}\}$$
$$I(F_5) = \{v_1\}$$
$$I(F_6) = \{v_2, e_{12}\}.$$

The reader may check that in this case $J(F_j) = I(F_j)$ for each j. The algorithm then proceeds inductively to define a discrete gradient on $\Delta(P)$. For F_1, the set $j_0 = \{v_0, e_{01}\}$ and the algorithm matches the vertex e_{01} in $\Delta(P)$ with the edge F_1, which corresponds to the chain $v_0 < e_{01}$. For $j = 2, \ldots, 5$, the algorithm matches the edge F_j with one of its vertices. Finally, for F_6, everything has already been matched except for F_6 itself, which remains unpaired. The discrete gradient is shown in Figure 9.2.

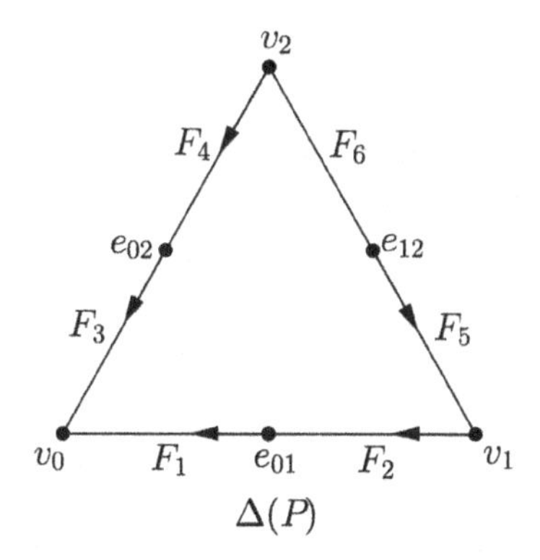

Fig. 9.2 The discrete gradient on $\Delta(P)$

Theorem 9.8. *The matching M on $\Delta(P)$ is acyclic.*

Proof. First assume that the I-intervals and the J-intervals agree. We show that M is acyclic on each $F_k - (\cup_{i<k} F_i)$ and that there are no directed cycles involving multiple pieces of the decomposition. Note that the second fact follows easily since any edge between distinct fibers of e is oriented from the later fiber to the earlier one. It remains that there can be no cycles in a single fiber. Call a cell *hollow* if it lacks the minimal rank vertex in the first interval where it differs from the critical cell. Call a cell *doubly-hit* if it includes the minimal rank vertex and at least one other vertex in the interval. The matching pairs hollow and doubly-hit cells with each other. Suppose then that we have an upward oriented edge $a \prec b$ in a cycle in the fiber. Then there must be a downward edge from b to some other c. Note that any cycle involves only two ranks of the face poset since we cannot have two upward edges in a matching. It follows that a and c are hollow and b is doubly-hit. Also, c is obtained from b by deleting some rank above the rank where a and b differ (if not, c would belong to an earlier orbit). We then see that the cycle must pass through cells that agree with the critical cell (if there is one) to higher and higher ranks; thus, it is impossible to complete the cycle.

Now, if the J-intervals differ from the I-intervals, we show that this matching on faces hitting all the J-intervals extends to $e^{-1}(F_k)$. The faces we need to consider are those faces in $F_k - (\cup_{i<k} F_i)$ that do not hit all the J-intervals (note they do hit the I-intervals). The algorithm guarantees that any rank r which gets truncated from an I-interval to make a J-interval j must belong to another I-interval at lower ranks. This lower I-interval gives rise to a J-interval, j'. Such an r cannot be the lowest rank in this I-interval, and a face which includes r will differ from the critical cell on j' and hence be matched by the algorithm, unless r is the lowest rank in j'. But this contradicts j being a minimal skipped interval since it would strictly contain j'. □

Theorem 9.9. *Each facet contributes at most one critical cell.*

Proof. If some node v is not in any minimal skipped interval of F_k then the minimal rank of v is used to match faces including v with those excluding v. It follows that $F_k - (\cup_{i<k} F_i)$ has no critical cells and the complex $F_1 \cup \cdots \cup F_k$ collapses to $F_1 \cup \cdots \cup F_{k-1}$. If every node in F_k belongs to at least one of its minimal skipped intervals, then the matching has a single critical cell consisting of the minimal ranks of the J-intervals. □

Corollary 9.10. *When F_k contributes a critical cell, the dimension of that cell is one less than the number of J-intervals for F_k.* □

Note that F_1 will always contribute a critical vertex since $I(F_1) = \emptyset$ and the algorithm leaves the lowest rank vertex unpaired.

Let Π_n be the poset of set partitions of $\{1, 2, \ldots, n\}$ ordered by refinement. Let $\lambda = (\lambda_1, \ldots, \lambda_k)$ be a partition of the integer n into unordered parts $\lambda_1 \geq \cdots \geq \lambda_k > 0$. Denote by S_λ the subgroup $S_{\lambda_1} \times \cdots \times S_{\lambda_k}$ of the symmetric group S_n. The group S_λ acts in a natural way on $\{1, 2, \ldots, n\}$ and therefore induces an order-preserving action on Π_n. Babson and Hersh use the matching described above to determine the structure of the quotient poset Π_n/S_λ for certain partitions λ. In particular, if λ is hook-shaped, then Π_n/S_λ is either collapsible or has the homotopy type of a wedge of spheres concentrated in top dimension. We refer the reader to [Babson and Hersh (2005)] for details.

9.2 Geographic data

Recall the topographical map of Pilot Mountain shown in Figure 1.2. We know that there are two obvious local maxima and a saddle between them (we can see them, after all), but suppose all we were given is a sampling of elevation values at some grid underlying the region. If we use the grid as vertices of a triangulation of the square, then we obtain a 2-dimensional simplicial complex X, and the elevation values give us a function f on the vertices of X. We may therefore use Algorithm 8.22 to construct a discrete gradient on X that mirrors the behavior of the function f. In this section we describe the output of this algorithm.

Take a 41×41 grid on the topographical map shown in Figure 1.2 and measure the elevation (in feet) at each point. The result of running Algorithm 8.22 on this data set is shown in Figure 9.3 (the figure shows the detail near the summit). Critical simplices are indicated by large dots for vertices, an edge with a dot at the midpoint for critical edges, and a dot with three short lines pointing to three regular vertices for critical triangles. The algorithm found both maxima and the saddle; the maximal triangles each have a vertex with the highest value among the points surrounding it, as one would expect.

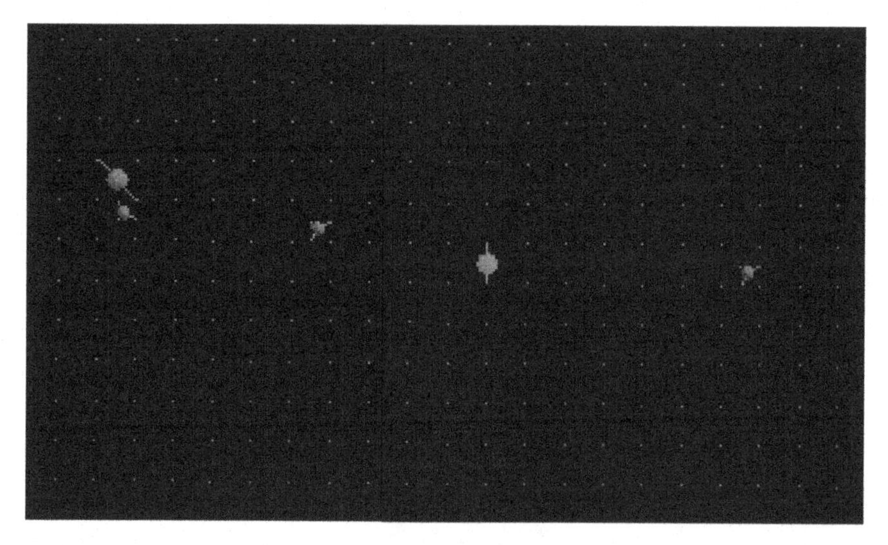

Fig. 9.3 The area around the summit. The absolute maximum is the triangle on the right; the saddle is the edge in the center; the local maximum is the triangle center left. The extra maximum and saddle are on the left side.

Also, the algorithm found an additional maximum and saddle in the left portion of the figure. The corresponding region of the topographical map shows that the area in question is relatively flat (indeed, it is the parking area at the summit), and estimating the elevation at each grid point is difficult. This is where the persistence parameter p comes into play. The elevation values at points in this area all lie in a very small range. By setting persistence to $p = 10$ (recall we are measuring elevation in feet), we can cancel these extra critical simplices while leaving the others intact. After implementing the cancellation routine, the extras disappear.

The algorithm also produces many critical vertices and edges along the boundary of the square (not shown). These are constrained minima and maxima on the boundary, which is topologically a circle.

This simple example illustrates the utility of Algorithm 8.22 in analyzing naturally occurring data. Other possible applications include meteorological data (temperature, humidity, air pressure, etc.), oceanographic data (temperature, salinity, etc.), and medical imaging data. For further examples, see [King et al. (2005)],[King et al. (2014)],[Robins et al. (2011)], etc. Also, geographical data has been analyzed via quasi MS-complexes; see [Edelsbrunner et al. (2003b)] for details.

Algorithms similar to Algorithm 8.22 have been presented by Robins, et al. [Robins et al. (2011)] and Lewiner, et al. [Lewiner et al. (2003c)] to analyze grayscale images. In particular, the algorithm in [Robins et al. (2011)] does not generate any additional critical cells beyond the minimal number, provided the dimension of the complex is at most 3. It is therefore perhaps a better alternative to Algorithm 8.22 in low dimensions.

9.3 Evasiveness

Recall the classical game *20 Questions.* This is a two-player game in which one person thinks of an object and the other must guess what it is by asking simple yes/no questions. For example, Person A may be thinking of a chair, and Person B then asks a sequence of questions: Is it alive? (No.) Is it bigger than a breadbox? (Yes.) Is it made of wood? (Yes.) and so on.

In this section we consider the following mathematical version of this game, analyzed by Forman in [Forman (2000)]. Let Σ be a simplex with vertices $v_0, v_1, \ldots, v_n$ and let M be a subcomplex of Σ which is known to both players. Person A chooses a face $\sigma \in \Sigma$ and Person B is allowed to ask questions of the type: Is vertex v_i in σ? Person B must then determine whether σ is in M in as few questions as possible. Assume further that Person B is allowed to use past information to choose the next question; we call this a *decision tree algorithm.*

Definition 9.11. If A is a decision tree algorithm, denote by $Q(\sigma, A, M)$ the number of questions Person B must ask to determine whether or not $\sigma \in M$. The *complexity* of M is

$$c(M) = \inf_A \sup_\sigma Q(\sigma, A, M).$$

Definition 9.12. The complex M is *evasive* if $c(M) = n + 1$ and *nonevasive* otherwise.

In other words, a complex is evasive if for any decision tree algorithm A there is a face $\sigma \in \Sigma$ such that Person B must ask $n+1$ questions to determine if σ is in M.

Definition 9.13. Let M be an evasive complex and let A be a decision tree algorithm. A simplex σ such that $Q(\sigma, A, M) = n + 1$ is called an *evader* of A.

Lemma 9.14. *Evaders occur in pairs* $\{\sigma_1 < \sigma_2\}$ *such that*

(1) $\dim \sigma_2 = \dim \sigma_1 + 1$;
(2) $\sigma_1 \in M$ *and* $\sigma_2 \notin M$.

Proof. Let A be a decision tree algorithm with $(n+1)$th question "Is v_i in σ?" and suppose α is an evader for A and $v_i \notin \alpha$. Then $\beta = v_i * \alpha$ is also an evader since Person B cannot distinguish between the possibilities $\sigma = \alpha$ (i.e., $\sigma \in M$) and $\sigma = \beta$ (i.e., $\sigma \notin M$) until $n+1$ questions have been asked. Similarly, if α is an evader with $v_i \in \alpha$, then $\gamma = \alpha - v_i$ is also an evader since γ and α are not distinguished until the $(n+1)$th question. □

Lemma 9.15. *A decision tree algorithm A induces a partition P of all the faces of Σ.*

Proof. Let P be the collection of all pairs $\{\alpha^{(p)} < \beta^{(p+1)}\}$ where α and β are not distinguished by A until the $(n+1)$th question. Then this is clearly a partition of the faces of Σ, including the empty face $\emptyset$ which is paired with some vertex v. Lemma 9.14 asserts that evaders are pairs $\{\alpha^{(p)} < \beta^{(p+1)}\}$ with $\alpha \in M$ and $\beta \notin M$. □

Theorem 9.16. *Let $V = P - \{\emptyset, v\}$. Then V is a discrete gradient vector field on Σ.*

Proof. Let α be a p-simplex in Σ and suppose $\sigma = \alpha$. Person B asks $n+1$ questions according to the algorithm A of the form "Is v_i in σ?" Number the questions from 1 to $n+1$ and let $n_0(\alpha) < n_1(\alpha) < \cdots < n_p(\alpha)$ be the numbers of the questions which receive the answer "yes." Denote the string $n_0(\alpha), \ldots, n_p(\alpha)$ by $n(\alpha)$. We will use these $n(\alpha)$ to define an ordering $\succ$ on the simplices of Σ.

Let $\alpha^{(p)}$ and $\beta^{(q)}$ be simplices of Σ and suppose first that $q > p$ and $n_i(\alpha) = n_i(\beta)$ for $i \leq p$ (we say that $n(\beta)$ is an *extension* of $n(\alpha)$). In this case set $\alpha \succ \beta$. In general, if neither $n(\alpha)$ nor $n(\beta)$ is an extension of the other, then order $n(\alpha)$ and $n(\beta)$ lexicographically and set $\alpha \succ \beta$ if there is a $k \leq \min\{p, q\}$ with $n_i(\alpha) = n_i(\beta)$ for $i < k$ and $n_k(\alpha) < n_k(\beta)$. This defines a transitive order on the simplices of Σ.

Now suppose $\alpha_0^{(p)} < \beta_0^{(p+1)} > \alpha_1^{(p)}$ is a portion of a V-path with $\alpha_0 \neq \alpha_1$. Since $\{\alpha_0 < \beta_0\} \in V$ we have

$$n_i(\beta_0) = n_i(\alpha_0), i \leq p,$$

and

$$n_{p+1}(\beta_0) = n+1;$$

that is, $n(\beta_0)$ is an extension of $n(\alpha_0)$ and hence $\alpha_0 \succ \beta_0$. Denote the vertices of β_0 by $u_0, u_1, \ldots, u_{p+1}$. Then if $\sigma = \alpha_0$ or $\sigma = \beta_0$, question number $n_i(\beta_0)$ is "Is u_i in σ?" Then the vertices of α_1 are $u_0, u_1, \ldots, u_{k-1}, u_{k+1}, \ldots, u_{p+1}$ for some $k \leq p$. If $\sigma = \beta_0$ then the first $n_k(\beta_0) - 1$ questions involve only $u_0, \ldots, u_{k-1}$ and vertices which are not in β_0 (and thus are also not in α_0, α_1). It follows that the first $n_k(\beta_0) - 1$ answers are the same whether $\sigma = \alpha_0, \beta_0$, or α_1. We therefore conclude that if $\sigma = \alpha_0, \beta_0$, or α_1, then question $n_k(\beta_0)$ is "Is u_k in σ?" If $\sigma = \beta_0$ the answer is "yes," and if $\sigma = \alpha_1$ the answer is "no." It follows that

$$n_i(\alpha_1) = n_i(\beta_0), i < k,$$

and

$$n_k(\alpha_1) > n_k(\beta_0),$$

from which we conclude that $\beta_0 \succ \alpha_1$.

So, if $\alpha_0 < \beta_0 > \alpha_1 < \beta_1 > \cdots < \beta_r > \alpha_{r+1}$ is any V-path, we have

$$\alpha_0 \succ \beta_0 \succ \alpha_1 \succ \cdots \succ \beta_r \succ \alpha_{r+1},$$

from which it follows that there are no closed V-paths. □

Theorem 9.16 allows us to deduce information about the topology of the complex M. Note that the empty set can be an evader for a decision tree algorithm. However, if M contains all the vertices of Σ then this is not possible.

Theorem 9.17. *Let A be a decision tree algorithm and let k be the number of pairs of evaders of A. Assume the empty set is not an evader of A. Then there is an ordering $\sigma_1, \sigma_2, \ldots, \sigma_k$ of the evaders of A which lie in M, an ordering $\tau_1, \tau_2, \ldots, \tau_k$ of the evaders of A which do not lie in M, and a nested sequence of subcomplexes of Σ*

$$\Sigma \supset \Sigma_k \supset \cdots \supset \Sigma_1 \supset \Sigma_0 = M \supset M_k \supset \cdots \supset M_1 \supset M_0 = v,$$

where v is a vertex of M which is not an evader of A, such that the following hold.

(1) Σ collapses onto Σ_k and M collapses onto M_k.
(2) For each $i = 1, \ldots, k$, τ_i is a maximal simplex of Σ_i not contained in Σ_{i-1}.
(3) For each $i = 1, \ldots, k$, σ_i is a maximal simplex of M_i not contained in M_{i-1}.
(4) For each $i = 1, \ldots, k$, $\Sigma_i - \tau_i$ collapses onto Σ_{i-1} and $M_i - \sigma_i$ collapses onto M_{i-1}.

If $\emptyset$ is an evader of A, the same result holds if we set $\sigma_1 = \emptyset$ and set $M_0 = M_1 = \emptyset$. In this case $M_2 = \sigma_2$ must be a vertex.

Proof. Consider the vector field V (Theorem 9.16). Let $W \subset V$ be the set of pairs $\{\alpha < \beta\} \in V$ such that either both α and β lie in M or neither lies in M. Then W is a discrete gradient and its critical simplices are described as follows. First note that if $\{\alpha < \beta\} \in V$ is such that $\alpha \subset M$ and $\beta \not\subset M$, then α and β are critical for W. Also, the vertex v, which is paired with $\emptyset$ in the partition P (Lemma 9.15) is critical for W. These are the only critical simplices for W since the only critical simplex of V is v. Note that v is an evader of A if and only if $\emptyset$ is an evader and this happens if and only if v is a vertex of M.

Let f be a discrete Morse function on Σ with gradient W. Note that if $\alpha^{(p)} \subset M$ and $\beta^{(p+1)} \not\subset M$ are such that $\beta > \alpha$ then $f(\beta) > f(\alpha)$. Consider the following quantities

$$\begin{aligned}
a &= \sup_{\alpha \subset M} f(\alpha) \\
b &= \inf_{\alpha \subset M} f(\alpha) \\
c &= 1 + a - b \\
d &= \inf_{\alpha \subset \Sigma} f(\alpha)
\end{aligned}$$

and define a new function g on Σ by

$$g(\alpha) = \begin{cases} f(\alpha) & \text{if } \alpha \subset M \\ f(\alpha) + c & \text{if } \alpha \not\subset M. \end{cases}$$

If $v \in M$ (i.e., $\emptyset$ is not an evader of A), set $g(v) = d - 1$ so that v is the global minimum of g. Then for each $\alpha \subset M$ and $\beta \not\subset M$ we have

$$g(\beta) \geq c + 1 > c \geq g(\alpha).$$

Also, for every pair $\alpha^{(p)} < \beta^{(p+1)}$ we have

$$g(\beta) > g(\alpha) \Leftrightarrow f(\beta) > f(\alpha).$$

It follows that g is a discrete Morse function on Σ with the same critical simplices as f; that is, the critical simplices of g are v and the evaders of A. The result now follows from Theorem 7.4 (see also Exercise 5). □

We note the following consequences of this result.

Corollary 9.18. *If $\{\sigma < \tau\}$ is a pair of evaders of A, define the index of such a pair to be* $\dim \sigma$. *Denote by $e_p(A)$ the number of pairs of evaders of A of index p. Here* $\dim \emptyset = -1$ *and $e_{-1} = 1$ if and only if $\emptyset$ is an evader of A. Then M is homotopy equivalent to a CW-complex with exactly $e_p(A)$ cells of dimension p, $p \geq 1$ and $e_0(A) - e_{-1}(A) + 1$ cells of dimension* 0.

Proof. This follows directly from Theorem 9.17 of course, but another proof may be obtained as follows. Let U be the restriction of the vector field V to M. The critical simplices of U are the evaders of A which lie in M and the vertex v (if it lies in M; i.e., $\emptyset$ is not an evader of A). Now apply Theorem 7.4. □

Given the complex M, denote by $\tilde{\beta}_i$ the reduced ith Betti number of M. Then we have the following result.

Corollary 9.19. *For any decision tree algorithm A,*

$$\#\{\text{pairs of evaders of } A\} \geq \sum_{i=0}^{n} \tilde{\beta}_i.$$

Corollary 9.20. *Let M be a nonempty subcomplex of Σ and suppose M is nonevasive. Then $\Sigma \searrow M$ and M is collapsible.*

Proof. Since there are no evaders, the lists of simplices σ_i and τ_i of Theorem 9.17 are empty; that is, $\Sigma_k = \Sigma_0 = M$ and $M = M_0 = v$ for any vertex v in M. □

Example 9.21. The complex of not-connected graphs, revisited. Recall Example 9.1, which concerned the complex X_n of not-connected graphs on n vertices. The complex X_n has the homotopy type of a wedge of $(n-1)!$ copies of the sphere S^{n-3} (Theorem 9.2). Let E be the set of possible edges of a graph on n vertices;

E has cardinality $\binom{n}{2}$. Let Σ be a simplex with $\binom{n}{2}$ vertices and let $N \subset \Sigma$ be the union of the faces of Σ corresponding to graphs which are not connected. Then X_n is clearly isomorphic to N. The question of whether or not a particular graph G is connected is then equivalent to asking if the face of Σ corresponding to G lies in N or not. We can decide this by inquiring about one vertex of Σ at a time. If A is a decision tree algorithm, then a graph G is an evader of A if we must examine all $\binom{n}{2}$ possible edges before determining if G is connected.

Proposition 9.22. *Let A be a decision tree algorithm for determining whether or not a graph on n vertices is connected. Then the number of evaders of A is at least* $2(n-1)!$.

Proof. By Corollary 9.19 we know that the number of pairs of evaders of A is the sum of the reduced Betti numbers of X_n. Theorem 9.2 tells us that this sum is $(n-1)!$ from which the result follows. □

9.4 Algebraic discrete Morse theory

One of the major applications of discrete Morse theory is the computation of the homology of a complex X. Since homology calculations arise in more general situations it is natural to ask if there is a corresponding theory for arbitrary chain complexes. This question has been addressed by several authors [Kozlov (2005)],[Sköldberg (2006)],[Batzies and Welker (2002)]; we present the version due to Kozlov [Kozlov (2005)].

Suppose A is an arbitrary commutative ring and C_* is a chain complex of free A-modules. We assume that C_* is bounded on the right. We choose a basis Ω_n of each C_n. If $\alpha \in C_n$ and $b \in \Omega_n$ denote the coefficient of b in the representation of α by $K_\Omega(\alpha, b)$.

Given $\Omega = \bigcup_n \Omega_n$, we obtain a ranked poset $P(C_*, \Omega)$ with A-weights on the order relations. If $b \in \Omega_n$ and $a \in \Omega_{n-1}$, the weight of the covering relation $b \succ a$ is $w_\Omega(b \succ a) = K_\Omega(\partial b, a)$.

Definition 9.23. A *partial matching* $\mathcal{M} \subseteq \Omega \times \Omega$ is a partial matching on the covering graph of $P(C_*, \Omega)$ such that if $b \succ a$ and $(a, b) \in \mathcal{M}$, then $w(b \succ a)$ is invertible in A.

Given such a matching, write $b = u(a)$ and $a = d(b)$ if $(a, b) \in \mathcal{M}$. Denote by $\mathcal{U}_n(\Omega)$ the set of all $b \in \Omega_n$ such that b is matched with some $a \in \Omega_{n-1}$ and by $\mathcal{D}_n(\Omega)$ the set of all $a \in \Omega_n$ that are matched with some $b \in \Omega_{n+1}$. The set $\mathcal{C}_n(\Omega) = \Omega_n \setminus (\mathcal{U}_n(\Omega) \cup \mathcal{D}_n(\Omega))$ consisting of all unmatched basis elements is called the set of *critical elements*.

Now, given two basis elements $s \in \Omega_n$ and $t \in \Omega_{n-1}$, an *alternating path* is a sequence

$$p = (s \succ d(b_1) \prec b_1 \succ d(b_2) \prec b_2 \succ \cdots \succ d(b_n) \prec b_n \succ t),$$

where $n \geq 0$ and all the $b_i \in \mathcal{U}(\Omega)$ are distinct. The *weight* of such a path is

$$w(p) = (-1)^n \frac{w(s \succ d(b_1))w(b_1 \succ d(b_2)) \cdots w(b_n \succ t)}{w(b_1 \succ d(b_1))w(b_2 \succ d(b_2)) \cdots w(b_n \succ d(b_n))}.$$

Definition 9.24. A partial matching on (C_*, Ω) is called *acyclic* if there does not exist a cycle

$$d(b_1) \prec b_1 \succ d(b_2) \prec b_2 \succ \cdots \succ d(b_n) \prec b_n \succ d(b_1),$$

with $n \geq 2$ and all b_i distinct.

Proposition 9.25. *A partial matching on (C_*, Ω) is acyclic if and only if there is a linear extension of $P(C_*, \Omega)$ such that $u(a)$ follows directly after a for all $a \in \mathcal{D}(\Omega)$.*

Proof. If L is such an extension, then if we follow an alternating path from left to right we always descend in the order; hence there can be no closed paths.

Conversely, suppose the matching is acyclic. Define L inductively as follows. Let Q be the set of elements already ordered in L, beginning with $Q = \emptyset$. Let W be the set of lowest rank elements of $P(C_*, \Omega) - Q$. Then one of the following cases holds.

Case 1. There is a critical element c in W. Then add c to L as the largest element and proceed with $Q \cup \{c\}$.

Case 2. All elements in W are matched. The covering graph of $W \cup u(W)$ is acyclic and so the total number of edges is at most $2|W| - 1$. But then there is an $a \in W$ with $P(C_*, \Omega)_{<u(a)} - Q = \{a\}$. Add $\{a, u(a)\}$ to the top of L and proceed with $Q \cup \{a, u(a)\}$. □

Definition 9.26. Let (C_*, Ω) be a free chain complex with basis, and let $\mathcal{M}$ be an acyclic matching. The *Morse complex*

$$\cdots \xrightarrow{\partial^{\mathcal{M}}_{n+2}} C^{\mathcal{M}}_{n+1} \xrightarrow{\partial^{\mathcal{M}}_{n+1}} C^{\mathcal{M}}_{n} \xrightarrow{\partial^{\mathcal{M}}_{n}} C^{\mathcal{M}}_{n-1} \xrightarrow{\partial^{\mathcal{M}}_{n-1}} \cdots$$

is defined as follows. The module $C^{\mathcal{M}}_n$ is the free A-module with basis $\mathcal{C}_n(\Omega)$. The boundary map is defined by

$$\partial^{\mathcal{M}}_n(s) = \sum_p w(p)p,$$

for $s \in C_n(\Omega)$ where the sum is taking over all alternating paths p with initial term s.

In the topological setting, we know that the discrete Morse complex recovers the homology of the space under consideration (Theorem 7.18). It is natural to believe, then, that the same is true in this more general algebraic setting. We now describe the exact result.

Definition 9.27. The chain complex

$$\cdots \to 0 \to A \overset{\mathrm{id}}{\to} A \to 0 \to \cdots$$

where the only nontrivial modules are in degrees d and $d-1$ is called an *atom chain complex*, denoted by $\mathtt{Atom}(d)$.

Theorem 9.28. *Let (C_*, Ω) be a free chain complex with basis and let $\mathcal{M}$ be an acyclic matching. Then C_* decomposes as a direct sum of chain complexes*

$$C_* \cong C_*^{\mathcal{M}} \oplus \bigoplus_{(a,b)\in\mathcal{M}} \mathtt{Atom}(\dim b).$$

Since the complex $T_* = \bigoplus_{(a,b)\in\mathcal{M}} \mathtt{Atom}(\dim b)$ is clearly acyclic, we have the following corollary.

Corollary 9.29. *The homology of C_* coincides with that of $C_*^{\mathcal{M}}$.* □

Proof of Theorem 9.28. Fix a linear extension L of the poset $P(C_*, \Omega)$ (Proposition 9.25) and denote the linear order by $<_L$. Assume first that C_* is bounded; without loss of generality we may assume $C_i = 0$ for $i < 0$ and $i > N$. Let $m = |\mathcal{M}|$ be the size of the matching and let $\ell = |\Omega| - 2m$ denote the number of critical elements. We will inductively build a sequence of bases $\Omega^0, \Omega^1, \ldots, \Omega^m$ of C_*, each of which will be partitioned as $\mathcal{C}(\Omega^k) = \{c_1^k, \cdots, c_\ell^k\}$, $\mathcal{D}(\Omega^k) = \{a_1^k, \ldots, a_m^k\}$, and $\mathcal{U}(\Omega^k) = \{b_1^k, \ldots, b_m^k\}$, such that $a_i^k = d(b_i^k)$ for $i = 1, \ldots, m$.

The initial step is to set $\Omega^0 = \Omega$ with the initial condition $b_i^0 <_L b_{i+1}^i$ for all $i = 1, \ldots, m-1$. The matching $\mathcal{M}$ gives the partition of Ω^0 into the pieces $\mathcal{C}, \mathcal{D}$, and $\mathcal{U}$. We now proceed inductively to construct the bases Ω^k while simultaneously proving the following two statements:

(1) $C_* = C_*[k] \oplus \mathcal{A}_1^k \oplus \cdots \oplus \mathcal{A}_k^k$, where $C_*[k]$ is the subcomplex generated by $\Omega^k - \{a_1^k, \ldots, a_k^k, b_1^k \ldots, b_k^k\}$ and $\mathcal{A}_i^k$ is the atom complex $\mathtt{Atom}(\dim b_i^k)$ for $i = 1, \ldots, k$.
(2) for each $x^k \in \mathcal{U}(\Omega^k) \cup \mathcal{C}(\Omega^k)$, $w(x^k \succ y^k) = \sum_p w(p)$, where the sum is restricted to alternating paths from x^0 to y^0 which use only the pairs (a_i^0, b_i^0) for $i = 1, \ldots, k$.

Clearly these are true for $k = 0$. Assume now that $k \geq 1$ and define the basis Ω^k from Ω^{k-1} as follows. Set $a_k^k = \partial b_k^{k-1}$, $b_k^k = b_k^{k-1}$, and $x^k = x^{k-1} - w(x^{k-1} \succ a_k^{k-1}) b_k^{k-1}$ for all $x^{k-1} \in \Omega^{k-1}$, $x^{k-1} \neq a_k, b_k$. To see that this is a basis, say $b_k^{k-1} \in C_n$. For $i \neq n, n-1$, we have $\Omega_i^k = \Omega_i^{k-1}$ and hence Ω_i^k is a basis of C_i by induction. The set Ω_{n-1}^k is constructed by adding a linear combination of basis elements to the element a_k^{k-1} and so Ω_{n-1}^k is a basis of C_{n-1}. The set Ω_n^k is obtained from Ω_n^{k-1} by subtracting multiples of b_k^{k-1} from other basis elements, so it is a basis of C_n.

We now determine the structure of the poset $P(C_*, \Omega^k)$. If $x \neq b_k$, then

$$\begin{aligned} w(x^k \succ a_k^k) &= K(\partial x^k, a_k^k) \\ &= K(\partial x^k, a_k^{k-1}) \\ &= K(\partial x^{k-1}, a_k^{k-1}) - w(x^{k-1} \succ a_k^{k-1}) K(\partial b_k^{k-1}, a_k^{k-1}) \\ &= 0. \end{aligned}$$

Moreover, since Ω_n^k is obtained from Ω_n^{k-1} by subtracting multiples of b_k^{k-1} from other basis elements, it follows that if $x \in \Omega_{n+1}^k$ and $y \in \Omega_n^k$, $y \neq b_k$ we have $w(x^k \succ y^k) = w(x^{k-1} \succ y^{k-1})$. Also, since the square of the differential vanishes we have

$$\begin{aligned} 0 &= \sum_{z^k \in \Omega_n^k} w(x^k \succ z^k) w(z^k \succ a^k) \\ &= w(x^k \succ b_k^k) w(b_k^k \succ a_k^k) \\ &= w(x^k \succ b_k^k), \end{aligned}$$

where the second equality follows from the fact that $w(z^k \succ a_k^k) = 0$ for all $z \neq b_k$. Thus, the weights in the poset $P(C_*, \Omega^k)$ are the same as in $P(C_*, \Omega^{k-1})$ except for the following cases:

(1) $w(x^k \succ b_k^k) = 0$ and $w(b_k^k \succ x^k) = 0$ for $x \neq a_k$;
(2) $w(a_k^k \succ x^k) = 0$ and $w(x^k \succ a_k^k) = 0$ for $x \neq b_k$;
(3) $w(x^k \succ y^k) = w(x^{k-1} \succ y^{k-1}) - w(x^{k-1} \succ a_k^{k-1}) w(b_k^{k-1} \succ y^{k-1})$ for $x \in \Omega_n^k$, $y \in \Omega_n^k$, $x \neq b_k$, $y \neq a_k$.

This proves the first statement above. Moreover, we claim that $(*)$ if $w(x^k \succ y^k) \neq w(x^{k-1} \succ y^{k-1})$ then $b_k^0 \geq_L y^0$. To see this, note that one of the following is true: (i) $y \in [a_k, b_k]$, (ii) y is critical, or (iii) $y = a_{k'}$ for some $k' > k$ such that $w(b_k^{k-1} \succ y^{k-1}) \neq 0$. In the first two cases, $b_k^0 \geq_L y^0$ by the construction of L. The last case is impossible by induction and by the construction of L. We then have $w(b_j^k \succ a_j^k) = w(b_j^{k-1} \succ a_j^{k-1})$ for all j and k; this is clear for $j = k$, the case $j < k$ follows by induction, and the case $j > k$ is a consequence of $(*)$.

Now define the partial matching $\mathcal{M}^k$ to be the set of pairs $\{(a_i^k, b_i^k) : i = 1, \ldots, m\}$. We claim this matching is acyclic. If $j \leq k$, the poset elements b_j^k and a_j^k are incomparable with the rest and hence cannot be part of a cycle. If $i > k$, $w(b_j^k \succ a_i^k) = w(b_j^{k-1} \succ a_i^{k-1})$ by $(*)$, and so, by induction, no cycle can be formed from these elements.

It remains to compute the boundary operator; that is, to prove the second statement above. Let $x^k \in \mathcal{U}(\Omega^k) \cup \mathcal{C}(\Omega^k)$, $y \in \mathcal{C}(\Omega^k)$. If $x^k = b_k$, the statement is clear. If $x^k \neq b_k$, then $w(x^k \succ y^k) = w(x^{k-1} \succ y^{k-1}) - w(x^{k-1} \succ a_k^{k-1}) w(b_k^{k-1} \succ y^{k-1})$. By induction, the first term contributes all the alternating paths from x^0 to

y^0 which use the edge $b_k^0 \succ a_k^0$. If this edge occurs, then by the construction of L, it must be the second edge of the path and by the fact $(*)$ we have $w(x^{k-1} \succ a_k^{k-1}) = w(x^0 \succ a_k^0)$. This proves the second statement and completes the proof under the assumption that C_* is bounded.

The unbounded case then follows immediately since the basis stabilizes in each dimension and we may take the union of these as the new basis for C_*. □

Note that Theorem 9.28 gives a new proof of Theorem 7.18. Indeed, a discrete Morse function induces a partial matching $\mathcal{M}$ on the chain complex $C_* = C_*(X, \mathbb{Z})$ and the complex $C_*^{\mathcal{M}}$ is the complex $\mathbb{M}_*(X, \mathbb{Z})$ (Theorem 7.19).

9.5 Homology reduction algorithm

Continuing the theme of Section 9.4, in this section we describe an algorithm, due to Harker, et al. [Harker et al. (2014)] to build a Morse complex from an arbitrary complex. In particular, this algorithm is useful for data analysis and computational dynamics, since it avoids the explicit construction of a global boundary operator, focusing instead on local information.

Let X be a CW complex (not necessarily regular) and let R be a PID of coefficients for homology. If $\sigma^{(p)} < \tau^{(p+1)}$ are cells in X, denote by $\kappa(\tau, \sigma)$ the R-incidence number of σ in $\partial\tau$; that is

$$\partial\tau = \sum_{\sigma^{(p)} < \tau} \kappa(\tau, \sigma)\sigma.$$

Definition 9.30. A *matching* of (X, κ) is a partition of X into three sets $\mathcal{A}$, $\mathcal{K}$, and $\mathcal{Q}$ and a bijection $w : \mathcal{Q} \to \mathcal{K}$ such that for each $Q \in \mathcal{Q}$ there is a unit $u \in R$ with $\kappa(w(Q), Q) = u$. Denote this decomposition by $(\mathcal{A}, w : \mathcal{Q} \to \mathcal{K})$.

Note that this definition is similar to the definition of an acyclic matching on the set of cells of a complex, or on an arbitrary chain complex. We will call the elements of $\mathcal{A}$ critical. The algorithms presented below will produce generators of the homology of X, and it is thus desirable to keep track of the elements in the sets $\mathcal{Q}$ and $\mathcal{K}$.

Now, given a matching of (X, κ), define a transitivity relation $\leq$ on $\mathcal{Q}$ as follows. Suppose $Q \neq Q' \in \mathcal{Q}$. If $\kappa(w(Q), Q') \neq 0$ then $Q' < Q$.

Definition 9.31. A matching of (X, κ) is *acyclic* if $\leq$ defines a partial order on $\mathcal{Q}$.

We now present an algorithm for constructing acyclic matchings on a complex.

Definition 9.32. Let $X' \subset X$ be a subcomplex. A pair of cells $(\tau, \tau') \in X' \times X'$ is a *coreduction pair* in X' if $\kappa(\tau, \tau')$ is a unit in R and $\partial_{X'}\tau = \{\tau'\}$.

Here, $\partial_{X'}\tau$ is the boundary of τ inside the subcomplex X'. In general, this may or may not equal the regular boundary of τ. Coreduction pairs are analogous to pairs of cells giving an elementary collapse in a complex (Definition 6.13).

Definition 9.33. Let $X' \subset X$ be a subcomplex. A cell $\tau \in X'$ is *free* in X' if $\partial_{X'}\tau = \emptyset$.

Algorithm 9.34. Coreduction-based Matching

1: Given X
2: $X' \longleftarrow X$
3: **while** X' is not empty **do**
4:     **while** X' admits a coreduction pair (K, Q) **do**
5:         Excise the pair (K, Q) from X'
6:         Add $K \in \mathcal{K}$ and $Q \in \mathcal{Q}$
7:         Set $w(Q) = K$
8:     **end while**
9:     **while** X' does not admit a coreduction pair **do**
10:         Excise a free cell A from X'
11:         Add $A \in \mathcal{A}$
12:     **end while**
13: **end while**
14: **return** $(\mathcal{A}, w : \mathcal{Q} \to \mathcal{K})$

Theorem 9.35. *Algorithm 9.34 produces an acyclic matching* $(\mathcal{A}, w : \mathcal{Q} \to \mathcal{K})$.

Proof. It is not clear, a priori, that the algorithm terminates. Note that each of the interior while statements reduces the size of the complex X'. At each stage, X' remains a subcomplex of X since excising a coreduction pair or a free cell yields a subcomplex. It is therefore sufficient to prove that as long as $X' \neq \emptyset$, one of these while statements can be executed. If a coreduction pair exists, then there is no problem, so assume there is not such a pair in X'. Since all complexes under consideration are finite, X' contains a cell τ of minimal dimension which must be free by definition. So the algorithm eventually terminates, resulting in a partition $\mathcal{A}$, $\mathcal{K}$, $\mathcal{Q}$. It is a matching since coreduction pairs satisfy the condition in Definition 9.30. It remains to show that this matching is acyclic.

Note that excising a free cell does not affect the partial order on $\mathcal{Q}$ defined above. If we add a new maximal element to a poset, then the partial order is unaffected below the new element. We claim that if the coreduction pair (K, Q) is excised, then Q is maximal in $\mathcal{Q}$. If not, there exists a previously excised coreduction pair (K', Q') with $Q \in \partial_{X'}K'$. But this is impossible: at that stage, call the subcomplex X'', the boundary $\partial_{X''}K'$ would have contained both Q and Q', violating the coreduction pair property of (K', Q') in X''. □

Suppose we have an acyclic matching $(\mathcal{A}, w : \mathcal{Q} \to \mathcal{K})$ on X. Denote by $C_\bullet(X)$

the chain complex on X (with coefficients in the ring R), and let $\langle \cdot, \cdot \rangle$ be the inner product on $C_\bullet(X)$ obtained by setting the cells to be an orthonomal basis.

Definition 9.36. The submodule $C_\bullet(\mathcal{K}) \subset C_\bullet(X)$ will be called the $\mathcal{K}$-*chains*. The submodule $C_\bullet(\mathcal{A}) \oplus C_\bullet(\mathcal{K})$ is called the module of *canonical chains.*

Our goal is to show that the module $C_\bullet(\mathcal{A})$ recovers the homology of X. As a first step, we will construct a chain map $\gamma_\bullet : C_\bullet(X) \to C_{\bullet+1}(X)$ with desirable properties. We present an algorithm.

Algorithm 9.37. Gamma Algorithm

1: **given** $x_{\text{in}} \in C_\bullet(X)$
2: $x \longleftarrow x_{\text{in}}$
3: $c \longleftarrow 0 \in C_\bullet(X)$
4: **while** $x \notin C_\bullet(\mathcal{A}) \oplus C_\bullet(\mathcal{K})$ **do**
5: choose a $\leq$-maximal $Q \in \mathcal{Q}$ with $\langle x, Q \rangle \neq 0$
6: $K \longleftarrow w(Q)$
7: $\omega \longleftarrow -\langle x, Q \rangle / \kappa(K, Q)$
8: $c \longleftarrow c + \omega K$
9: $x \longleftarrow x + \omega \partial K$
10: **end while**
11: **return** c

Proposition 9.38. *Let $x \in C_\bullet(X)$. Using x as the input for Algorithm 9.37 yields a $\mathcal{K}$-chain c such that $x + \partial c$ is canonical. Moreover, if x is canonical then $c = 0$.*

Proof. Let $\{x^i\}$ and $\{c^i\}$ be the sequences of values of x and c produced by the algorithm. It is easy to see inductively that for each i we have $x^i = x_{\text{in}} + \partial c^i$. We first show that the algorithm terminates. Let Q_i denote the maximal element chosen via x^i. A direct calculation shows that $\langle x^{i+1}, Q_i \rangle = 0$. Since Q_i is maximal, we see that $\langle x^{i+j}, Q_i \rangle = 0$ for $j \geq 1$. Since the set $\{Q \in \mathcal{Q} : \langle x, Q \rangle \neq 0\}$ is finite, there exists a j such that

$$\{Q \in \mathcal{Q} : \langle x^j, Q \rangle \neq 0\} = \emptyset,$$

in which case x^j is canonical (by definition) and the algorithm stops. □

Proposition 9.39. *Let x be a $\mathcal{K}$-chain. If ∂x is canonical, then $x = 0$.*

Proof. Suppose $x \neq 0$. Choose $K \in \mathcal{K}$ with $Q = w^{-1}(K)$ $\leq$-maximal in the set

$$\{w^{-1}(K') \in \mathcal{Q} : \langle x, K' \rangle \neq 0\}.$$

We know $\kappa(K, Q) \neq 0$. By assumption, ∂x is canonical and thus $\langle \partial x, Q \rangle = 0$. Since R has no zero divisors, there exists a $K' \neq K$ with $\langle x, K' \rangle \neq 0$ and $\kappa(K', Q) \neq 0$. But then $Q \leq w^{-1}(K)$, contrary to the maximality of Q. □

Proposition 9.40. *The module homomorphism* $\gamma_\bullet : C_\bullet(X) \to C_{\bullet+1}(X)$ *is the unique map with the following properties:*

(1) the image of $\mathrm{id}_X + \partial\gamma$ *is canonical;*
(2) for each $x \in C_\bullet(X)$, $\gamma(x)$ *is a* $\mathcal{K}$*-chain.*

Moreover, $\ker\gamma = C_\bullet(\mathcal{A}) \oplus C_\bullet(\mathcal{K})$.

Proof. Proposition 9.38 has taken care of all but the uniqueness. Suppose that for $x \in C_\bullet(X)$ there are $\mathcal{K}$-chains c_1, c_2 such that $x + \partial c_i$ is canonical. Then $c_1 - c_2$ is a $\mathcal{K}$-chain and $\partial(c_1 - c_2)$ is canonical. Proposition 9.39 then implies $c_1 = c_2$. □

Proposition 9.41. *The sequence*

$$C_\bullet(X) \overset{\mathrm{id}_X + \partial\gamma}{\longrightarrow} C_\bullet(X) \overset{\gamma}{\longrightarrow} C_{\bullet+1} \overset{\mathrm{id}_X + \gamma\partial}{\longrightarrow} C_{\bullet+1}(X)$$

is exact. In particular, $\ker\gamma$ *is the canonical submodule and* $im\,\gamma$ *is the submodule of* $\mathcal{K}$*-chains.*

Proof. Proposition 9.40 implies that $\ker\gamma$ and $\mathrm{im}(\mathrm{id}_X + \partial\gamma)$ equal the submodule of canonical chains. We need only check that $\ker(\mathrm{id}_X + \gamma\partial) = \mathrm{im}\,\gamma$ is the submodule of $\mathcal{K}$-chains. Let x be a $\mathcal{K}$-chain. By Proposition 9.40 $\gamma\partial x$ is the unique $\mathcal{K}$-chain such that $\partial x + \partial\gamma\partial x$ is canonical. Since 0 is canonical and $-x$ is a $\mathcal{K}$-chain, we see that $\gamma\partial x = -x$. It follows that x lies in $\mathrm{im}\,\gamma$ and also that $(\mathrm{id}_X + \gamma\partial)x = 0$. Conversely, if $x \in \ker(\mathrm{id}_X + \gamma\partial)$, then $x + k = 0$ for some $\mathcal{K}$-chain k and so x is a $\mathcal{K}$-chain. □

Now, denote by $i_{\mathcal{A}} : C_\bullet(\mathcal{A}) \to C_\bullet(X)$ the inclusion morphism and by $\pi_{\mathcal{A}} : C_\bullet(X) \to C_\bullet(\mathcal{A})$ the projection map. Define module maps $\psi : C_\bullet(X) \to C_\bullet(\mathcal{A})$ and $\phi : C_\bullet(\mathcal{A}) \to C_\bullet(X)$ by

$$\psi = \pi_{\mathcal{A}} \circ (\mathrm{id}_X + \partial\gamma) \text{ and } \phi = (\mathrm{id}_X + \gamma\partial) \circ i_{\mathcal{A}},$$

and define $\Delta : C_\bullet(\mathcal{A}) \to C_{\bullet-1}(\mathcal{A})$ by

$$\Delta = \psi \circ \partial \circ \phi.$$

Lemma 9.42. $\phi \circ \psi = \mathrm{id}_X + \partial\gamma + \gamma\partial$ *on* $C_\bullet(X)$.

Proof. We have the following:

$$\phi \circ \psi = (\mathrm{id}_X + \gamma\partial) \circ i_{\mathcal{A}} \circ \pi_{\mathcal{A}} \circ (\mathrm{id}_X + \partial\gamma).$$

By Proposition 9.41, $i_{\mathcal{A}} \circ \pi_{\mathcal{A}} \circ (\mathrm{id}_X + \partial\gamma)$ differs from $(\mathrm{id}_X + \partial\gamma)$ by a $\mathcal{K}$-chain k, which gets mapped to 0 by $\pi_{\mathcal{A}}$. Since k is in the kernel of $\mathrm{id}_X = \gamma\partial$, we see that $\phi \circ \psi = (\mathrm{id}_X + \gamma\partial) \circ (\mathrm{id}_X + \partial\gamma)$. Since $\partial^2 = 0$, the result follows. □

Theorem 9.43. $(C_\bullet(\mathcal{A}), \Delta)$ *is a chain complex, and* $\psi : C_\bullet(X) \to C_\bullet(\mathcal{A})$ *and* $\phi : C_\bullet(\mathcal{A}) \to C_\bullet(X)$ *are chain equivalences.*

Proof. We first show that $\Delta \circ \Delta = 0$. By definition

$$\Delta \circ \Delta = (\psi \circ \partial \circ \phi) \circ (\psi \circ \partial \circ \phi).$$

Lemma 9.42 asserts that the composition $\phi \circ \psi$ is $(\mathrm{id}_X + \partial\gamma + \gamma\partial)$. But then we have

$$\begin{aligned}\partial \circ (\phi \circ \psi) \circ \partial &= \partial \circ (\mathrm{id}_X + \partial\gamma + \gamma\partial) \circ \partial \\ &= \partial \mathrm{id}_X \partial + \partial\partial\gamma\partial + \partial\gamma\partial\partial \\ &= 0.\end{aligned}$$

It follows that $\Delta \circ \Delta = \psi \circ 0 \circ \phi = 0$.

We now show that ϕ and ψ are chain maps. That is, we must show that $\Delta \circ \phi = \phi \circ \Delta$ and similarly for ψ. We have, by definition,

$$\begin{aligned}\phi \circ \Delta &= \phi \circ (\phi \circ \partial \circ \phi) \\ &= (\mathrm{id}_X + \partial\gamma + \gamma\partial) \circ \partial \circ \phi \text{ (by Lemma 9.42)} \\ &= (\mathrm{id}_X + \partial\gamma) \circ \partial \circ \phi \text{ (since } \partial^2 = 0) \\ &= (\mathrm{id}_X + \partial\gamma) \circ \partial \circ (\mathrm{id}_X + \gamma\partial) \circ i_{\mathcal{A}}.\end{aligned}$$

But by Proposition 9.41, we know that $\ker \gamma$ consists of the canonical chains $\mathcal{A}$ and hence this final composition equals $\partial \circ \phi$. A similar argument shows that $\Delta \circ \psi = \psi \circ \partial$.

It remains to show that the maps ϕ and ψ are chain equivalences. Lemma 9.42 shows that $\phi \circ \psi$ is homotopic to id_X. We claim that $\psi \circ \phi = \mathrm{id}_{\mathcal{A}}$. Note that

$$\psi \circ \phi = \pi_{\mathcal{A}} \circ (\mathrm{id}_X + \partial\gamma) \circ (\mathrm{id}_X + \gamma\partial) \circ i_{\mathcal{A}}.$$

But $\mathrm{im}\, \gamma = C_\bullet(\mathcal{K}) \subset C_\bullet(\mathcal{K}) \oplus C_\bullet(\mathcal{A}) = \ker \gamma$ and so $\psi \circ \phi = \pi_{\mathcal{A}} \circ \mathrm{id}_X \circ i_{\mathcal{A}} = \mathrm{id}_{\mathcal{A}}$. □

Theorem 9.43 shows that the module built on the critical cells of an acyclic matching, with the map Δ, is indeed a chain complex and that its homology agrees with that of X. On its face, this may not seem to be an improvement over the algorithms of the previous chapter. However, since one often wishes to work with complexes that arise from numerical or experimental datasets, we see that we have made an improvement: we can iterate Algorithm 9.34 to reduce the size of the complex further. There is no real analogue of this if one wishes to define discrete gradient fields.

Moreover, it is possible to keep track of homology generators. This is often useful in applications.

Algorithm 9.44. Homology Generators Algorithm

1: **given** a complex (X, ∂) (or (X, κ))
2: Use Algorithm 9.34 to produce an acyclic matching $(\mathcal{A}, w : \mathcal{Q} \to \mathcal{K})$
3: **for all** $A \in \mathcal{A}$ **do**
4: compute and store ΔA using Algorithm 9.37 and the definition of Δ
5: **end for**

6: **if** $\mathcal{A} = X$ **then**
7: Use Smith Normal Form to compute homology generators $\{g_i\}$ of $(C_\bullet(\mathcal{A}), \Delta)$
8: **end if**
9: **return** $\{\phi(g_i)\}$

This approach is also useful for computing the induced map on homology associated to a map of complex. We refer the reader to [Harker et al. (2014)] for further details. Also, these algorithms are relatively efficient. To be more precise, let us define the *complex mass* of (X, κ) as follows:

$$m = \#\{(\sigma, \tau) \in X^2 : \kappa(\sigma, \tau) \neq 0\}.$$

This is a better measure of the complex; indeed, it gives a count of the number of incidences, which in general is much smaller than the worst case estimate obtainable from the number n of cells of X.

Proposition 9.45. *Algorithm 9.34 applied to (X, κ) executes in $O(m)$ time and requires $O(n)$ memory.*

Proof. For details, we refer the reader to [Harker et al. (2014)], Proposition 5.1. □

9.6 Exercises

(1) Let X be the 2-skeleton of a standard 3-simplex. X is homeomorphic to the 2-sphere and geometrically is the surface of a tetrahedron. Apply Algorithm 9.7 to construct a Morse matching on X.
(2) Download some geographic data (e.g, from `http://www.ngdc.noaa.gov/mgg/topo/globe.html`) and use the implementation of Algorithm 8.22 available online to generate a discrete Morse function from the data, similar to the Pilot Mountain example of Section 9.2.
(3) A discrete Morse function on a simplicial complex X induces a partial matching $\mathcal{M}$ on the chain complex $C_* = C_*(X, \mathbb{Z})$. Prove that the complex $C_*^{\mathcal{M}}$ of Section 9.4 is the discrete Morse complex $\mathbb{M}_*(X, \mathbb{Z})$; that is, check that the differentials in these complexes agree.
(4) Take any simplicial complex (a triangulated torus, for example), and apply Algorithm 9.34 to produce an acyclic matching on the complex. Then use Algorithm 9.44 to produce a collection of generators for the homology of the complex.

Appendix A

Smooth Manifolds

In this appendix, we collect the basics of smooth manifolds that are used in the text. Proofs of these results may be found in any textbook on manifolds, for example [Munkres (1963)].

Basic definitions

An *m-dimensional manifold* is a second countable Hausdorff space M such that each point $x \in M$ has an open neighborhood U which is homeomorphic to an open ball B about the origin in $\mathbb{R}^m$. We call the homeomorphism $\phi_U : U \to B$ a *coordinate chart* at x and usually assume that $\phi_U(x) = 0$. These charts are required to satisfy a compatibility condition: if U and V are coordinate neighborhoods in M, with $U \cap V \neq \emptyset$ then the map $\tau_{U,V} : \phi_U(U \cap V) \to \phi_V(U \cap V)$ defined by $\tau_{U,V} = \phi_V \circ \phi_U^{-1}$ is a homeomorphism. The manifold is of class C^∞ if each of the transition maps $\tau_{U,V}$ is of class C^∞ (i.e., it has derivatives of all orders). We often refer to such a manifold as *smooth.*

Definition A.1. Let M be a smooth manifold. A function $f : M \to \mathbb{R}$ is *smooth* if at each point $x \in M$ it is of class C^∞ with respect to a local coordinate system; that is, if $\phi_U^{-1} : B \to U$ is a smooth coordinate system at x, then the composition $f \circ \phi_U^{-1} : B \to \mathbb{R}$ is of class C^∞.

More generally, if M and N are smooth manifolds, a function $f : M \to N$ is *smooth* if it is smooth locally: if $x \in M$, choose coordinate neighborhoods U of x and V of $f(x)$ so that $f(U) \subset V$, and then insist that the map f is C^∞ in these coordinate systems (i.e., when viewed as a map $\mathbb{R}^m \to \mathbb{R}^n$, $n = \dim N$).

Recall that a point $x \in M$ is a *critical point* of $f : M \to \mathbb{R}$ if

$$\frac{\partial f}{\partial x_i}(x) = 0, \quad i = 1, \ldots, m,$$

where $(x_1, \ldots, x_m)$ is a local coordinate system around x. The real number $f(x)$ is then called a *critical value* of f.

In this book, we often work with manifolds with boundary. The canonical example of such an object is the closed disc $D^m \subset \mathbb{R}^m$ consisting of points of length ≤ 1. The bounding sphere S^{m-1} is an $(m-1)$-dimensional manifold contained in D^m and each point $x \in S^{m-1}$ has a coordinate neighborhood diffeomorphic to the upper half disc $U = \{(x_1, \ldots, x_m) : x_m \geq 0\}$. In general, we call M an *m-manifold with boundary* if each point $x \in M$ has a coordinate neighborhood homeomorphic to an open disc in $\mathbb{R}^m$ or the upper half disc U. Points in the latter class form the *boundary of M*, denoted ∂M. In fact, it is possible to show that a manifold with boundary M has the following property: there is a smooth function $f : M \to \mathbb{R}$ such that 0 is not a critical value of f with the property that $\partial M = f^{-1}(0)$. For example, if we define $f : D^m \to \mathbb{R}$ by

$$f(x_1, \ldots, x_m) = 1 - x_1^2 + \cdots + x_m^2$$

then $S^{m-1} = f^{-1}(0)$. The implicit function theorem guarantees that if $f : M \to \mathbb{R}$ is a smooth map such that 0 is not a critical value, then $f^{-1}(0)$ is an $(m-1)$-submanifold of M.

It should be clear how to define the notion of a smooth map $f : M \to N$ between manifolds with boundary. In the study of Morse theory, we use the following gluing construction.

Theorem A.2. *Let M and N be manifolds with boundary and let $\varphi : \partial M \to \partial N$ be a diffeomorphism. Then the space $W = M \cup_\varphi N$ obtained by identifying $x \in \partial M$ with $\varphi(x) \in \partial N$ has the structure of a smooth manifold, and it is unique up to diffeomorphism.* □

Gluing diffeomorphisms together is more subtle. Suppose $\varphi : \partial M_1 \to \partial M_2$ and $\psi : \partial N_1 \to \partial N_2$ are diffeomorphisms and let $W = M_1 \cup_\varphi M_2$, $V = N_1 \cup_\psi N_2$.

Theorem A.3. *Suppose $h_1 : M_1 \to N_1$ and $h_2 : M_2 \to N_2$ are diffeomorphisms such that $\psi \circ h_1|_{\partial M_1} = h_2 \circ \varphi|_{\partial M_1}$. Then there is a diffeomorphism $h : W \to V$ such that $h = h_1$ on $M_1 - \partial M_1$ and $h = h_2$ on $N_1 - \partial N_1$.* □

The proof of this is a bit tricky since one needs to be careful about the definition of h near the boundaries of M_1 and N_1. The trick is to modify h_1 and h_2 near the boundaries by a small perturbation to make them match nicely.

Tangent vectors

Suppose M is an m-manifold embedded in $\mathbb{R}^n$ and that p is a point on M. Recall that a *tangent vector* is a vector based at p which is tangent to M; the set of all such vectors is called the *tangent space of M at p*, denoted $T_p(M)$. In the case of a surface ($m = 2$) embedded in $\mathbb{R}^3$ this is the familiar tangent plane from introductory calculus. The set $T_p(M)$ carries the structure of a vector space of dimension m.

The most common example of a tangent vector is provided by the velocity vector of a curve. Suppose $\sigma : (a,b) \to \mathbb{R}^n$ is a smooth curve. Write the coordinates of $\mathbb{R}^n$ as $(x_1, x_2, \ldots, x_n)$ and let t be the parameter for σ. Then we may write σ in coordinates:

$$\sigma(t) = (x_1(t), x_2(t), \ldots, x_n(t)), a < t < b.$$

We may as well assume that $0 \in (a,b)$ and that $\sigma(0) = p$. The velocity vector $\mathbf{v}$ of the curve at $t = 0$ is

$$\mathbf{v} = \frac{d\sigma}{dt}(0) = \left(\frac{dx_1}{dt}(0), \frac{dx_2}{dt}(0), \ldots, \frac{dx_n}{dt}(0)\right).$$

If σ happens to lie in M, then the velocity vector $\mathbf{v}$ is a tangent vector to M at p.

Recall the notion of *directional derivative* from calculus. Given a curve σ in M, a tangent vector $\mathbf{v} = (v_1, v_2, \ldots, v_n) \in T_p(M)$ with $\frac{d\sigma}{dt}(0) = \mathbf{v}$, and a smooth function f defined in a neighborhood of p in $\mathbb{R}^n$, the derivative of f in the direction of $\mathbf{v}$ is

$$\begin{aligned}\left.\frac{df(\sigma(t))}{dt}\right|_{t=0} &= \left.\frac{d}{dt}f(x_1(t), x_2(t), \ldots, x_n(t))\right|_{t=0} \\ &= \sum_{i=1}^{n} \frac{\partial f}{\partial x_i}(p)\frac{dx_i}{dt}(0) \\ &= \sum_{i=1}^{n} v_i \frac{\partial f}{\partial x_i}(p).\end{aligned}$$

Note that this quantity depends only on f and $\mathbf{v}$; the curve σ does not matter. We may therefore write this as

$$\mathbf{v} \cdot f.$$

Observe that $\mathbf{v} \cdot f > 0$ precisely when the function $f(\sigma(t))$ is an increasing function of t near $t = 0$ and so the directional derivative is positive if and only if $\mathbf{v}$ points in the direction where f is increasing.

Let us now recall how to construct a basis of $T_p(M)$. Let $(x_1, x_2, \ldots, x_m)$ be a coordinate system about $p = (p_1, p_2, \ldots, p_m) \in M$ and define a curve $\sigma_i(t)$ in M by

$$\sigma_i(t) = (p_1, \ldots, p_{i-1}, p_i + t, p_{i+1}, \ldots, p_m).$$

This curve has unit speed in the direction x_i and the corresponding velocity vector at p is denoted by $\mathbf{e}_i$. The collection

$$\mathbf{e}_1, \mathbf{e}_2, \ldots, \mathbf{e}_m$$

forms a basis for $T_p(M)$. Note that if f is a smooth function on M, the derivative of f in the direction of $\mathbf{e}_i$ is

$$\begin{aligned}\mathbf{e}_i \cdot f &= \left.\frac{d}{dt}f(\sigma_i(t))\right|_{t=0} \\ &= \left.\frac{d}{dt}f(p_1, \ldots, p_{i-1}, p_i + t, p_{i+1}, \ldots, p_m)\right|_{t=0} \\ &= \frac{\partial f}{\partial x_i}(p).\end{aligned}$$

We often write

$$\mathbf{e}_i = \left(\frac{\partial}{\partial x_i}\right)_p.$$

Example A.4. Let M be the unit sphere in $\mathbb{R}^3$ and consider the point $p = (1/\sqrt{2}, 1/\sqrt{2}, 1/2)$ on M. We may parametrize M using spherical coordinates (φ, θ); the point p in these coordinates is $(\pi/4, \pi/4)$. The basis vectors are easily seen to be $\mathbf{e}_1 = (1, 0)$ and $\mathbf{e}_2 = (0, 1)$. It is instructive to realize these as embedded in $\mathbb{R}^3$, however. In spherical coordinates, the curves σ_i are as follows

$$\sigma_1(t) = \left(\cos\left(\frac{\pi}{4}\right)\sin\left(\frac{\pi}{4}+t\right), \sin\left(\frac{\pi}{4}\right)\sin\left(\frac{\pi}{4}+t\right), \cos\left(\frac{\pi}{4}+t\right)\right)$$

$$\sigma_2(t) = \left(\sin\left(\frac{\pi}{4}\right)\cos\left(\frac{\pi}{4}+t\right), \sin\left(\frac{\pi}{4}\right)\sin\left(\frac{\pi}{4}+t\right), \cos\left(\frac{\pi}{4}\right)\right).$$

Differentiating these at $t = 0$ yields two basis vectors for the $T_p(M)$:

$$\mathbf{e}_1 = \left(\frac{1}{2}, \frac{1}{2}, -\frac{1}{\sqrt{2}}\right)$$

$$\mathbf{e}_2 = \left(-\frac{1}{2}, \frac{1}{2}, 0\right).$$

Note that these make good sense geometrically: the curve σ_1 is a piece of the great circle passing through the north pole and p, heading down the sphere, while the curve σ_2 is the circle of latitude passing through p, heading counterclockwise when viewed from above.

Consider the height function $f : M \to \mathbb{R}$ given by $f(x, y, z) = z$. In our coordinate system, we have $f(\varphi, \theta) = \cos\varphi$. Let us compute the derivative of f in the direction of $\mathbf{v} = (v_1, v_2)$ at p:

$$\begin{aligned}\mathbf{v} \cdot f &= v_1 \frac{\partial f}{\partial \varphi}(p) + v_2 \frac{\partial f}{\partial \theta}(p) \\ &= -v_1 \sin\left(\frac{\pi}{4}\right) \\ &= -\frac{v_1}{\sqrt{2}}.\end{aligned}$$

Note that this is positive precisely when $v_1 < 0$, which corresponds to a vector pointing toward the north pole when written in terms of the basis $\mathbf{e}_1, \mathbf{e}_2$.

It is clear that if two tangent vectors $\mathbf{u}$ and $\mathbf{v}$ satisfy

$$\mathbf{u} \cdot f = \mathbf{v} \cdot f$$

for *every* smooth function f, then $\mathbf{u} = \mathbf{v}$. Suppose $\sigma(t)$ is a curve in M passing through p when $t = 0$ and $(x_1, \ldots, x_m)$ is a local coordinate system centered at p.

Denote by $\mathbf{v}$ the velocity vector of σ at p. Write $\sigma(t) = (x_1(t), \ldots, x_m(t))$. If f is defined in a neighborhood of p then

$$\begin{aligned}
\mathbf{v} \cdot f &= \frac{d}{dt} f(\sigma(t)) \bigg|_{t=0} \\
&= \frac{d}{dt} f(x_1(t), \ldots, x_m(t)) \bigg|_{t=0} \\
&= \sum_{i=1}^{m} \frac{\partial f}{\partial x_i}(p) \frac{dx_i}{dt}(0) \\
&= \sum_{i=1}^{m} \frac{dx_i}{dt}(0) \mathbf{e}_i \cdot f,
\end{aligned}$$

and so

$$\mathbf{v} = \sum_{i=1}^{m} \frac{dx_i}{dt}(0) \left(\frac{\partial}{\partial x_i} \right)_p .$$

Moreover, it is possible to derive a formula for the coordinate transformation between two systems of coordinate vectors. If $(x_1, \ldots, x_m)$ and $(y_1, \ldots, y_m)$ are coordinate systems at p, then

$$\left(\frac{\partial}{\partial x_i} \right)_p = \sum_{j=1}^{m} \frac{\partial y_j}{\partial x_i}(p) \left(\frac{\partial}{\partial y_j} \right)_p .$$

Riemannian metrics

The *tangent bundle* of M is

$$TM = \{(x, \mathbf{v}) : \mathbf{v} \in T_x M\}.$$

It should be fairly obvious that TM is itself a manifold of dimension $2 \dim M$. There is an obvious map $\pi : TM \to M$ defined by $\pi(x, \mathbf{v}) = x$; the fiber over $x \in M$ is $T_x M$.

Definition A.5. A *Riemannian metric* on M is a family $\langle , \rangle_x$ of (positive definite) inner products

$$\langle , \rangle_x : T_x M \times T_x M \to \mathbb{R}$$

such that if X and Y are smooth vector fields on M, the map

$$x \mapsto \langle X(x), Y(x) \rangle_x$$

is a smooth map $M \to \mathbb{R}$.

A Riemannian metric on M allows one to move concepts defined at a single tangent space to adjoining regions in the manifold. We call it a metric since it allows us to define lengths of curves in the manifold M: if $\sigma : [0, 1] \to M$ is a curve, then if we have a Riemannian metric we can define the length of the tangent vector $||\sigma'(t)|| = \langle\sigma'(t), \sigma'(t)\rangle_{\sigma(t)}^{1/2}$ for any $t \in [0,1]$ in the tangent space $T_{\sigma(t)}M$. We then get the length of the curve as the integral

$$L(\sigma) = \int_0^1 ||\sigma'(t)||\, dt.$$

Orientation of a manifold

Let U be an open m-disc in $\mathbb{R}^m$. The tangent space T_xU is clearly just $\mathbb{R}^m$ itself. In particular, we may choose a collection of vector fields

$$V = \langle \mathbf{v}_1, \mathbf{v}_2, \ldots, \mathbf{v}_m \rangle$$

on U such that at each point $x \in U$, the collection $\{\mathbf{v}_i(x)\}$ forms an *ordered* basis of T_xU. We call such a collection V an *orientation* of U. If we have a similar collection W, then we say that V and W determine the same orientation of U if the matrix A transforming V to W has positive determinant at every point of U. (Note that if the determinant is positive at one point it is positive at all points.) Otherwise we say that V and W determine opposite orientations.

Definition A.6. Let M be an m-manifold. Cover M by a collection $\{U_\lambda\}_{\lambda\in\Lambda}$ of coordinate neighborhoods, each of which has an orientation V_λ. For each nonempty intersection $U_\lambda \cap U_\mu$, if the orientations V_λ and V_μ agree on the overlap, then we say M is *orientable* and call the V_λ an *orientation* of M, denoted by $\langle M \rangle$.

If M is orientable and connected, then there are exactly two orientations on M, opposite to each other.

Example A.7. The sphere S^m is orientable for all $m \geq 0$. Odd-dimensional projective spaces $\mathbb{R}P^{2k+1}$ are orientable, while even-dimensional projective spaces $\mathbb{R}P^{2k}$ are not. The torus is orientable; the Klein bottle is not.

Appendix B

Cell Complexes

In this appendix, we collect basic facts about cell complexes and simplicial complexes. A more comprehensive reference for the concepts used in the text is [Kozlov (2008)].

Abstract simplicial complexes

Definition B.1. An *abstract simplicial complex* is a finite set X together with a collection Δ of subsets of X such that if $A \in \Delta$ and $B \subset A$, then $B \in \Delta$.

We often denote the simplicial complex by Δ. If $v \in X$ is an element with $\{v\} \in \Delta$, we call v a *vertex* of Δ. Denote the set of vertices of Δ by $V(\Delta)$. If a collection $v_0, v_1, \ldots, v_i$ of vertices is such that $\{v_0, \ldots, v_i\} \in \Delta$, we call this set an *$i$-simplex* of Δ and denote it by $\langle v_0, v_1, \ldots, v_i \rangle$. The $(i-1)$-simplex obtained from removing the jth vertex from $\langle v_0, v_1, \ldots, v_i \rangle$ is called a *(codimension-1)-face* and we denote the removal of v_j by $\langle v_0, \ldots, v_{j-1}, \hat{v}_j, v_{j+1}, \ldots, v_i \rangle$.

Example B.2. Let $m \geq 0$ and consider the simplicial complex Δ^m whose vertex set is $X = \{v_0, \ldots, v_m\}$ and whose simplices are all the nonempty subsets of X. The complex Δ^m is called the *standard m-simplex.*

Suppose now that we have two simplicial complexes Δ_1 and Δ_2. A *simplicial map* from Δ_1 to Δ_2 is a map $f : V(\Delta_1) \to V(\Delta_2)$ such that if $\sigma \in \Delta_1$ is a simplex then $f(\sigma)$ is a simplex in Δ_2. In this case we simply write $f : \Delta_1 \to \Delta_2$. The collection of simplicial complexes and simplicial maps form a category, since the composition of two simplicial maps is clearly a simplicial map, as is the identity map on the set of vertices.

If we have a simplex τ in Δ, we may wish to delete it. We need to be careful, however, to delete all the simplices containing τ as a face as well. We therefore define the complex $\Delta - \tau$ to be the simplicial complex whose simplices are $\{\sigma \in \Delta : \sigma \not\supseteq \tau\}$.

We now consider analogues of neighborhoods of points in a manifold in this setting.

Definition B.3. Let σ be a simplex in Δ. The *link* of σ is the set

$$\mathrm{Lk}_{\Delta}(\sigma) = \{\tau \in \Delta : \sigma \cap \tau = \emptyset \text{ and } \sigma \cup \tau \in \Delta\}.$$

The *closed star* of σ is the subcomplex of Δ defined by

$$\mathrm{St}_{\Delta}(\sigma) = \{\tau \in \Delta : \tau \cup \sigma \in \Delta\}.$$

The *open star* is the set of simplices containing σ:

$$S_{\Delta}(\sigma) = \{\tau \in \Delta : \sigma \subseteq \tau\}.$$

Note that the link of σ is the intersection $\mathrm{St}(\sigma) \cap (\Delta - V(\sigma))$. The open star is the analogue of an open neighborhood around a point of the manifold, the closed star the closure of that neighborhood, and the link the boundary. To make the last assertion precise, we make the following definition.

Definition B.4. Let Δ_1 and Δ_2 be abstract simplicial complexes with disjoint vertex sets. The *join* of Δ_1 and Δ_2 is the abstract simplicial complex $\Delta_1 * \Delta_2$ whose vertex set is $V(\Delta_1) \cup V(\Delta_2)$ and whose simplices are all $\sigma \subseteq V(\Delta_1) \cup V(\Delta_2)$ such that

$$\sigma \cap V(\Delta_1) \in \Delta_1 \text{ and } \sigma \cap V(\Delta_2) \in \Delta_2.$$

With this definition, it is easy to see that if $\sigma \in \Delta$, then

$$\mathrm{St}_{\Delta}(\sigma) = \mathrm{Lk}_{\Delta}(\sigma) * \sigma.$$

We often need to subdivide simplicial complexes. The following two constructions are used in the text.

Definition B.5. Let Δ be an abstract simplicial complex. The *barycentric subdivision* of Δ is the abstract simplicial complex $\mathrm{sd}(\Delta)$ defined by

$$\mathrm{sd}(\Delta) = \{\{\sigma_1, \dots, \sigma_r\} : \sigma_1 \supset \sigma_2 \supset \cdots \supset \sigma_r, \sigma_i \in \Delta\} \cup \{\emptyset\}.$$

Note in particular that the vertices of $\mathrm{sd}(\Delta)$ are indexed by the nonempty simplices of Δ.

In some situations, we need only subdivide Δ at an individual simplex.

Definition B.6. Let σ be a simplex of Δ. The *stellar subdivision* of Δ at σ is the complex $\mathrm{sd}_{\Delta}(\sigma)$ defined as follows. Take the set of vertices to be $V(\Delta) \cup \{\hat{\sigma}\}$, where $\hat{\sigma}$ is a new vertex indexed by σ (if σ is itself a vertex then $\hat{\sigma} = \sigma$ and we have done nothing). A simplex $\tau \in \Delta$ is a simplex of $\mathrm{sd}_{\Delta}(\sigma)$ if and only if τ does not contain σ. We add new simplices of the form $\tau \cup \{\hat{\sigma}\}$ for $\tau \in \Delta$ such that $\tau \cup \sigma \in \Delta$ and $\tau \not\supset \sigma$.

Geometric realization

Abstract simplicial complexes did not come first historically. All of this theory is grounded in geometry.

Definition B.7. A *geometric n-simplex* σ is the convex hull of a set A of $n+1$ affinely independent points in $\mathbb{R}^N$ for some $N \geq n$. The convex hulls of the subsets of A are called the *faces* of σ. The *standard n-simplex* is the convex hull Δ^n of the set of endpoints of the standard basis vectors $(1, 0, \ldots, 0), (0, 1, 0, \ldots, 0), \ldots, (0, \ldots, 0, 1)$ in $\mathbb{R}^{n+1}$.

We may now define the geometric realization of an arbitrary finite abstract simplicial complex. First note that if we have a finite set A, we have a finite dimensional vector space $\mathbb{R}^A$ with basis elements corresponding to the elements of A. Subsets of A then correspond to subspaces of $\mathbb{R}^A$, and we can build geometric simplices in this vector space.

Definition B.8. Let Δ be a finite abstract simplicial complex. The *standard geometric realization* of Δ is the topological space obtained by taking the union of standard σ-simplices in $\mathbb{R}^{V(\Delta)}$ for all $\sigma \in \Delta$. Any topological space that is homeomorphic to the standard geometric realization is called a *geometric realization* of Δ and is denoted by $|\Delta|$.

We often blur the distinction between a simplicial complex Δ and its realization $|\Delta|$, especially when we are talking about topological properties.

Example B.9. Consider the simplicial complex Δ with $V(\Delta) = \{1, 2, 3\}$ and simplices $\{\{1, 2\}, \{1, 3\}, \{2, 3\}\}$. Of course, we think of this complex as a triangle, often embedded in the plane, but its standard geometric realization lives in $\mathbb{R}^3$ and is the triangle joining the points $(1, 0, 0), (0, 1, 0)$, and $(0, 0, 1)$. As a topological space, however, we often take $|\Delta| = S^1$, the unit circle in the plane.

We may also construct the geometric realization X of Δ via an inductive process. Begin with the vertex set $V(\Delta)$ and add simplices one by one in any order as long as we add simplices only when all their proper faces have already been added to the complex. Formally, if σ is a simplex, all of whose faces are in X, we build the space $X \cup_{\partial \Delta^\sigma} \Delta^\sigma$, where we identify the boundary of the geometric simplex Δ^σ corresponding to σ in $\mathbb{R}^{V(\Delta)}$ with the subspace of X corresponding to the proper faces of σ.

It should be clear that given a simplicial map $f : \Delta_1 \to \Delta_2$, we have an induced map $|f| : |\Delta_1| \to |\Delta_2|$ and that this map is continuous. In fact, it is the restriction of a linear map $\mathbb{R}^{V(\Delta_1)} \to \mathbb{R}^{V(\Delta_2)}$.

Consider the standard geometric n-simplex Δ^n. The barycentric subdivision of Δ^n has the following geometric interpretation. First consider the edges $\langle v_i, v_j \rangle$ for vertices $v_i \neq v_j$. We take the midpoint of this edge. This is then the stellar subdivision of $\langle v_i, v_j \rangle$. Given a 2-simplex $\sigma = \langle v_i, v_j, v_k \rangle$, the stellar subdivision of σ is obtained by taking the centroid of σ as $\hat{\sigma}$ and then taking as simplices the joins $\hat{\sigma} * \tau$ for each τ in the subdivisions of the various edges $\langle v_\ell, v_m \rangle$. We illustrate this in Figure B.1.

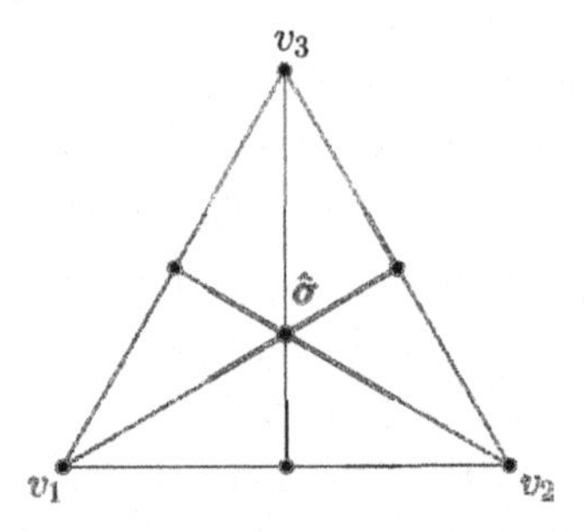

Fig. B.1 The barycentric subdivision of a 2-simplex

It should now be clear how one may view the barycentric subdivision of a general n-simplex Δ^n: inductively subdivide its faces in this manner. The stellar subdivision $\mathrm{sd}_\Delta(\sigma)$ is then the result of barycentrically subdividing σ and the barycentric subdivision $\mathrm{sd}(\Delta)$ is obtained by taking the stellar subdivision of each of its simplices.

Note that while the abstract simplicial complexes Δ and $\mathrm{sd}(\Delta)$ are not the same, their geometric realizations are:

$$|\Delta| \cong |\mathrm{sd}(\Delta)|.$$

We leave this as an exercise for the reader.

We might also take the following point of view. Since we have standard geometric simplices of any finite dimension, we could try to build simplicial complexes from these as subspaces of some ambient Euclidean space.

Definition B.10. A *geometric simplicial complex* K in $\mathbb{R}^N$ is a collection of geometric simplices in $\mathbb{R}^N$ such that every face of a simplex in K is a simplex of K and the intersection of any two simplices in K is a face of each of them. We denote the subcomplex of all simplices of dimension $\leq d$ by $K^{(d)}$ and call it the d-skeleton of K. Letting $|K|$ denote the union of all simplices in K, the topology on $|K|$ is induced from that of $\mathbb{R}^N$; that is, each simplex has the induced subspace topology and a set $U \subseteq |K|$ is open (resp., closed) if and only if the intersection $A \cap \sigma$ is open (resp., closed) for every simplex σ in K.

These two concepts are equivalent in the following sense. Given an abstract simplicial complex, its geometric realization is a geometric simplicial complex. Conversely, the combinatorial structure of the simplices of a geometric simplicial complex is an abstract simplicial complex.

Finally, we have the notion of polyhedral complex, which is used occasionally in the text. Recall that *convex polytope* P is a bounded subset of $\mathbb{R}^d$ that is the solution of a finite number of linear equalities and inequalities. A subset $F \subseteq P$ is called a *face* of P if there is a linear function f on $\mathbb{R}^d$ such that $f(s) = 0$ for all $s \in F$ and $f(p) \geq 0$ for all $p \in P$.

Definition B.11. A *geometric polyhedral complex* Γ in $\mathbb{R}^N$ is a collection of convex polytopes in $\mathbb{R}^N$ such that

(1) every face of a polytope in Γ is a polytope in Γ;
(2) the intersection of any two polytopes in Γ is a face of each of them.

Geometric polyhedral complexes generalize the notion of geometric simplicial complex. Indeed, any such Γ can be subdivided in a suitable way to obtain a simplicial complex: divide each edge in half, then for each face F take the barycenter b_F of F and cone to the simplices in its boundary. Examples of geometric polyhedral complexes that are not simplicial complexes as given include the cube and dodecahedron. A polyhedral complex whose cells are cubes of various dimensions is called a *cubical complex*.

CW complexes

By an *m-cell* we mean a topological space homeomorphic to the standard closed unit ball D^m in $\mathbb{R}^m$. An *open m-cell* is a space homeomorphic to the interior of D^m. We now describe a procedure to build spaces, called *CW-complexes*, with m-cells as building blocks.

Definition B.12. Let X and Y be topological spaces, let $A \subseteq X$ be a subspace, and let $f : A \to Y$ be a continuous map. Define the space $Y \cup_f X$ as the quotient space $X \coprod Y/\sim$, where $\sim$ is the equivalence relation generated by $a \sim f(a)$ for all $a \in A$. We say that the space $Y \cup_f X$ is obtained from Y by attaching X along f.

Note that the map f need not be injective. Also, the space $Y \cup_f X$ inherits the quotient topology from the disjoint union $X \coprod Y$.

We have already seen examples of this in the case of simplicial complexes. We attach a simplex σ along a continuous map $f : \partial\sigma \to Y$. This observation will imply that simplicial complexes are special cases of the more general construction we now describe.

Definition B.13. A *CW-complex* is a topological space X constructed via the following inductive procedure.

(1) The 0-skeleton $X^{(0)}$ is a discrete set.

(2) Assume the $(n-1)$-skeleton has been defined and construct the n-skeleton $X^{(n)}$ by attaching n-cells to $X^{(n-1)}$ along their boundaries. Give $X^{(n)}$ the associated quotient topology.
(3) The space $X = \bigcup_{n=0}^{\infty} X^{(n)}$ is given the weak topology: a set $A \subseteq X$ is open if and only if each $A \cap X^{(n)}$ is open in $X^{(n)}$.

We often refer to the attaching map $\chi_\alpha : e_\alpha^m \to X$ as the *characteristic map* of the cell. There are obvious notions of subcomplex, cellular map, etc., which we leave to the reader.

Definition B.14. A CW-complex X is *regular* if for each cell e_α the restriction of the characteristic map $\chi_\alpha : \partial e_\alpha \to \chi_\alpha(\partial e_\alpha)$ is a homeomorphism.

Much of the discrete Morse theory presented in the text may be extended to arbitrary CW-complexes even though we usually present the results for regular complexes.

Triangulations of manifolds

Finally, we mention the following definition.

Definition B.15. Let M be a topological m-manifold. A *triangulation* of M is a simplicial complex T together with a homeomorphism $f : |T| \to M$. A triangulation T is *piecewise linear* (PL, for short) if the link of any simplex in T is a PL sphere (that is, a triangulated sphere with an atlas of piecewise linear functions).

Any smooth manifold supports a triangulation (and an essentially unique PL-triangulation). In low dimensions (≤ 3), any manifold is triangulable. There are manifolds in dimensions ≥ 4, however, that cannot be triangulated. In this text, we mostly work with smooth manifolds, but in any case, we usually assume our manifolds have triangulations when necessary.

Bibliography

Adiprasito, K., Benedetti, B. (2013). Subdivisions, shellability, and collapsibility of products, arXiv:1202.6606.

Babson, E., Hersh, P. (2005). Discrete Morse functions from lexicographic orders, *Trans. Amer. Math. Soc.* **357**, 509–534.

Banchoff, T. (1970). Critical points and curvature for embedded polyhedral surfaces, *Amer. Math. Monthly* **77**, 475–485.

Bang–Jensen, J., Gutin, G. (2008). *Digraphs: Theory, Algorithms and Applications*, 2nd edn. (Springer–Verlag, London).

Batzies, E., Welker, V. (2002). Discrete Morse theory for cellular resolutions, *J. Reine Angew. Math.* **543**, 147–168.

Benedetti, B. (2012). Discrete Morse theory for manifolds with boundary, *Trans. Amer. Math. Soc.* **364**, 6631–6670.

Benedetti, B. (2014). Smoothing discrete Morse theory, arXiv:1212.0885.

Bestvina, M. (2008). PL Morse theory, *Math. Commun.* **13**, 149–162.

Bestvina, M., Brady, N. (1997). Morse theory and finiteness properties of groups, *Invent. Math.* **129**, 445–470.

Chari, M. (2000). On discrete Morse functions and combinatorial decompositions, *Formal Power Series and Algebraic Combinatorics, Discrete Math.* **217**, 101–113.

Edelsbrunner, H., Harer, J., Natarajan, V., Pascucci, V. (2003a). Morse-Smale complexes for piecewise linear 3-manifolds., *Proc. 19th Ann. Sympos. Comput. Geom.*, 361–370.

Edelsbrunner, H., Harer, J., Zomorodian, A. (2003b). Hierarchical Morse–Smale complexes for piecewise linear 2-manifolds, *ACM Symposium on Computational Geometry (Medford, MA, 2001), Discrete Comput. Geom.* **30**, 87–107.

Edelsbrunner, H., Harer, J. (2009) *Computational Topology, An Introduction*, (American Mathematical Society, Providence).

Eğecioglu, O., Gonzalez, T. (1996). A computationally intractable problem on simplicial complexes, *Comput. Geom.* **6**, 85–98.

Forman, R. (1998a). Morse theory for cell complexes, *Adv. Math.* **134**, 90–145.

Forman, R. (1998b). Combinatorial vector fields and dynamical systems, *Math. Z.* **228** (1998), 629–681.

Forman, R. (1998c). Witten–Morse theory for cell complexes, *Topology*, **37**, 945–979.

Forman, R. (2000). Morse theory and evasiveness, *Combinatorica* **20**, 489–504.

Forman, R. (2002). A user's guide to discrete Morse theory, *Sém. Lothar. Combin.* **48**, Art. B48c, 35pp.

Gallais, E. (2010). Combinatorial realization of the Thom–Smale complex via discrete Morse theory, *Ann. Sc. Norm. Super. Pisa Cl. Sci. (5)* **9**, 229–252.

Harker, S., Mischaikow, K., Mrozek, M., Nanda, V. (2014). Discrete Morse theoretic algorithms for computing homology of complexes and maps, *Found. Comput. Math.* **14**, 151–184.

Jonsson, J. (2008). *Simplicial Complexes of Graphs, Lecture Notes in Mathematics* Vol. 1928, (Springer–Verlag, Berlin).

Joswig, M., Pfetsch, M. (2006). Computing optimal Morse matchings, *SIAM J. Discrete Math.* **20**, 11–25.

Kearton, C., Lickorish, W.B.R. (1972). Piecewise linear critical levels and collapsing, *Trans. Amer. Math. Soc.* **170**, 415–424.

King, H., Knudson, K., Mramor, N. (2005). Generating discrete Morse functions from point data, *Experiment. Math.* **14**, 435–444.

King, H., Knudson, K., Mramor, N. (2014). Birth and death in discrete Morse theory, arXiv:0808.0051.

Kosiński, A. (1962). Singularities of piecewise linear mappings I, *Bull. Amer. Math. Soc.* **68**, 110–114.

Kozlov, D. (2005). Discrete Morse theory for free chain complexes, *C. R. Math. Acad. Sci. Paris* **340**, 867–872.

Kozlov, D. (2008) *Combinatorial Algebraic Topology, Algorithms and Computations in Mathematics* Vol. 21, (Springer, Berlin).

Lang, S. (2004). *Linear Algebra*, 3rd edn. (Springer, Berlin).

Lewiner, T., Lopes, H., Tavares, G. (2003a). Optimal discrete Morse functions for 2-manifolds, *Comput. Geom.* **26**, 221–233.

Lewiner, T., Lopes, H., Tavares, G. (2003b). Toward optimality in discrete Morse theory, *Experiment. Math.* **12**, 271–285.

Lewiner, T., Lopes, H., Tavares, G. (2003c). Visualizing Forman's discrete vector field, *Visualization and Mathematics III, Math. Vis.* (Springer, Berlin), 95–112.

Massey, W. (1991). *A Basic Course in Algebraic Topology*, (Springer–Verlag, New York).

Matsumoto, Y. (1997). *An Introduction to Morse Theory, Translations of Mathematical Monographs*, Vol. 208, (American Mathematical Society, Providence).

Milnor, J. (1956). On manifolds homeomorphic to the 7-sphere, *Ann. of Math. (2)* **64**, 399–405.

Milnor, J. (1963) *Morse Theory, Ann. of Math. Studies*, Vol. 51, (Princeton University Press, Princeton).

Milnor, J. (1965). *Topology from the Differentiable Viewpoint*, (University Press of Virginia Charlottesville).

Milnor, J. (1965). *Lectures on the h-cobordism Theorem*, (Princeton University Press)

Munkres, J. (1963). *Elementary Differential Topology, Ann. of Math. Studies* Vol. 54 (Princeton University Press).

Munkres, J. (1984). *Elements of Algebraic Topology*, (Addison–Wesley, Menlo Park, CA).

Robins, V., Wood, P., Sheppard, A. (2011) Theory and algorithms for constructing discrete Morse complexes from grayscale digital images, *IEEE Trans. Pattern Analysis and Machine Intelligence* **33**, 1646–1658.

Schwarz, M. (1993). *Morse Homology, Progress in Mathematics*, Vol. 111, (Birkhäuser Verlag, Basel).

Schwarz, M. (1999). Equivalences for Morse homology, *Geometry and Topology in Dynamics (Winston–Salem, NC, 1998/San Antonio, TX, 1999), Contemp. Math.*, Vol. 246, (American Mathematical Society, Providence), 197–216.

Shareshian, J. (2001) Discrete Morse theory for complexes of 2-connected graphs, *Topology* **40**, 681–701.

Sköldberg, E. (2006). Morse theory from an algebraic viewpoint, *Trans. Amer. Math. Soc.*

358, 115–129.

Stallings, J. (1968). *Lectures on Polyhedral Toplogy*, (Tata Institute of Fundamental Research, Bombay).

Vassiliev, V. (1993). Complexes of connected graphs, *The Gelfand Mathematical Seminars, 1990–1992*, (Birkhäuser–Boston, Boston), 223–235.

Index

0-handle, 28

ascending link, 80
attaching map, 32
attaching sphere, 32

barycentric subdivision, 170
belt sphere, 62
Betti number, 52

(C^2, ε)-approximation, 17
$\mathcal{C}(p, q)$, 56
canceling critical cells, 109
canonical chains, 158
chain complex with basis, 152
  Morse complex of, 153
closed star, 170
co-core, 29
cohomology, 61
complex mass, 161
complex of not-connected graphs, 141
convex polytope, 173
core, 29
coreduction pair, 156
critical cell of a discrete Morse function, 87
critical point, 5, 163
  index of, 15
  nondegenerate, 7
critical points
  isolated, 15
critical value, 18
cubical complex, 173
CW-complex, 173
  regular, 85, 174

decision tree algorithm, 148
  complexity of, 148
deformation retract, 28
descending link, 80
directional derivative, 22, 165
discrete gradient vector field, 93
discrete vector field, 92
  basic set of, 100
  chain recurrent set of, 100
  dual, 96
  Morse numbers of, 101
  refinement of, 97
dual complex, 96
dual pseudograph, 126

endocollapsible complex, 121
Euler characteristic, 53
  of an odd dimensional manifold, 54
evasive complex, 148

facet, 143
flow line, 55
free cell, 157

general projection map, 68
geometric realization, 171
gradient
  of a smooth function, 21
gradient flow, 55
  discrete, 112
gradient path
  multiplicity of, 117

handle cancellation, 45
handle slide, 43
handlebody, 32

handlebody decomposition, 32
  of $\mathbb{R}P^m$, 34
  of S^m, 34
  of $SO(m)$, 36
  of $U(m)$, 38
Hasse diagram, 101
  modified, 102
height function
  on S^2, 7
  on S^n, 16
Hessian, 6
homology, 51
  depends on coefficient ring, 53
  of $\mathbb{R}P^2$, 53
  of S^n, 52
hypercircuit, 128
hyperedge, 128
hyperforest, 128
hypergraph, 128
  associated bipartite graph, 128
  regular component of, 128

index, 15
  on a polyhedral surface, 68
integral curve, 24
isotopy, 40
  of an attaching map, 40

join, 170
junction, 74

k-fold saddle, 72
$\mathcal{K}$-chains, 158

lexicographic order, 143
link, 170
lower disc, 42
lower link, 131
lower star, 72
Lyapunov function, 101

m-handle, 28
  of index i, 29
manifold, 163
  orientation of, 168
  with corners, 56
matching on a complex, 156
middle vertex, 69
minimal skipped interval, 143
monkey saddle, 69

Morse complex, 54
  discrete, 111
    boundary map, 116
    homology of, 115
Morse function, 7
  discrete, 86
    from point data, 131
    on a 1-dimensional complex, 126
    on a 2-dimensional manifold, 126
    optimal, 136
  existence of, 17
  on a simplicial complex, 79
  perfect, 60
  self-indexing, 45, 58
  with two critical points, 16
Morse homology, 57
  agrees with cellular homology, 58
Morse inequalities
  strong, 59, 107
  weak, 59, 107
Morse Lemma, 13
Morse–Smale complex, 55
  combinatorial, 70

negative critical point, 77
NP-complete, 136
NP-hard, 135

open star, 170
orientation of a manifold, 168

Φ-invariant chains, 113
partial matching, 102
persistence pairing, 77
piecewise linear function, 67
PL-manifold, 174
Poincaré Conjcture, 108
Poincaré duality, 60
  for discrete Morse theory, 100
Poincaré polynomial, 59
polyhedral surface, 68
polyhedron, 173
positive critical point, 77
pure simplicial complex, 121

quadrangulation, 70
  splittable, 71
quasi Morse–Smale complex, 72, 73

regular face, 86

Riemannian metric, 167

shelling, 121
simplicial collapse, 90, 105, 108
simplicial complex, 169
simplicial map, 169
skeleton, 172
smoothed manifold, 29
stable manifold, 42, 55
stellar subdivision, 170
sublevel complex, 105
sublevel set, 9, 27, 29

tangent bundle, 167
tangent space, 164
tangent vector, 164
Tits building, 81
torus, height function on, 9
transverse intersection, 45
triangulation of a manifold, 174

unstable manifold, 42, 55
  orientation of, 56
upper disc, 42
upper star, 72

V-path, 94
  closed, 94
  non-trivial, 94
  refinement of, 97
vector field, 21
  gradient-like, 23

weight of a covering relation, 152

Printed in the United States
By Bookmasters